Africa's Power Infrastructure

Africa's Power Infrastructure

Investment, Integration, Efficiency

Anton Eberhard
Orvika Rosnes
Maria Shkaratan
Haakon Vennemo

Vivien Foster and Cecilia Briceño-Garmendia,
Series Editors

THE WORLD BANK
Washington, D.C.

ISBN: 978-0-8213-8455-8
eISBN: 978-0-8213-8652-1
DOI: 10.1596/978-0-8213-8455-8

Library of Congress Cataloging-in-Publication Data
Africa's power infrastructure : investment, integration, efficiency / Anton Eberhard ... [et al.].
  p. cm.
Includes bibliographical references.
ISBN 978-0-8213-8455-8 — ISBN 978-0-8213-8652-1 (electronic)
1. Rural electrification—Government policy—Africa, Sub-Saharan. 2. Energy policy—Social aspects—Africa, Sub-Saharan. 3. Capital investments—Africa, Sub-Saharan. I. Eberhard, Anton A.
  HD9688.S832.A37 2011
  333.793'20967—dc22

2011002973

Cover photograph: Arne Hoel/The World Bank
Cover design: Naylor Design

Contents

Boxes

Figures

About the AICD

This study is a product of the Africa Infrastructure Country Diagnostic (AICD), a project designed to expand the world's knowledge of physical infrastructure in Africa. The AICD provides a baseline against which future improvements in infrastructure services can be measured, making it possible to monitor the results achieved from donor support. It also offers a more solid empirical foundation for prioritizing investments and designing policy reforms in the infrastructure sectors in Africa.

The AICD was based on an unprecedented effort to collect detailed economic and technical data on the infrastructure sectors in Africa. The project produced a series of original reports on public expenditure, spending needs, and sector performance in each of the main infrastructure sectors, including energy, information and communication technologies, irrigation, transport, and water and sanitation. The most significant findings were synthesized in a flagship report titled *Africa's Infrastructure: A Time for Transformation*. All the underlying data and models are available to the public through a Web portal (http://www.infrastructureafrica.org), allowing users to download customized data reports and perform various simulation exercises.

The AICD was commissioned by the Infrastructure Consortium for Africa following the 2005 G-8 Summit at Gleneagles, which flagged the importance of scaling up donor finance to infrastructure in support of Africa's development.

The first phase of the AICD focused on 24 countries that together account for 85 percent of the gross domestic product, population, and infrastructure aid flows of Sub-Saharan Africa. The countries were Benin, Burkina Faso, Cape Verde, Cameroon, Chad, Democratic Republic of Congo, Côte d'Ivoire,

Ethiopia, Ghana, Kenya, Lesotho, Madagascar, Malawi, Mozambique, Namibia, Niger, Nigeria, Rwanda, Senegal, South Africa, Sudan, Tanzania, Uganda, and Zambia. Under a second phase of the project, coverage was expanded to include the remaining countries on the African continent.

Consistent with the genesis of the project, the main focus was on the 48 countries south of the Sahara that face the most severe infrastructure challenges. Some components of the study also covered North African countries to provide a broader point of reference. Unless otherwise stated, therefore, the term "Africa" is used throughout this report as a shorthand for "Sub-Saharan Africa."

The AICD was implemented by the World Bank on behalf of a steering committee that represents the African Union, the New Partnership for Africa's Development (NEPAD), Africa's regional economic communities, the African Development Bank, and major infrastructure donors. Financing for the AICD was provided by a multidonor trust fund to which the main contributors were the Department for International Development (United Kingdom), the Public Private Infrastructure Advisory Facility, Agence Française de Développement, the European Commission, and Germany's Kreditanstalt für Wiederaufbau (KfW). The Sub-Saharan Africa Transport Policy Program and the Water and Sanitation Program provided technical support on data collection and analysis pertaining to their respective sectors. A group of distinguished peer reviewers from policy-making and academic circles in Africa and beyond reviewed all of the major outputs of the study to ensure the technical quality of the work.

Following the completion of the AICD project, long-term responsibility for ongoing collection and analysis of African infrastructure statistics was transferred to the African Development Bank under the Africa Infrastructure Knowledge Program (AIKP). A second wave of data collection of the infrastructure indicators analyzed in this volume was initiated in 2011.

Series Foreword

The Africa Infrastructure Country Diagnostic (AICD) has produced continent-wide analysis of many aspects of Africa's infrastructure challenge. The main findings were synthesized in a flagship report titled *Africa's Infrastructure: A Time for Transformation*, published in November 2009. Meant for policy makers, that report necessarily focused on the high-level conclusions. It attracted widespread media coverage feeding directly into discussions at the 2009 African Union Commission Heads of State Summit on Infrastructure.

Although the flagship report served a valuable role in highlighting the main findings of the project, it could not do full justice to the richness of the data collected and technical analysis undertaken. There was clearly a need to make this more detailed material available to a wider audience of infrastructure practitioners. Hence the idea of producing four technical monographs, such as this one, to provide detailed results on each of the major infrastructure sectors—information and communication technologies (ICT), power, transport, and water—as companions to the flagship report.

These technical volumes are intended as reference books on each of the infrastructure sectors. They cover all aspects of the AICD project relevant to each sector, including sector performance, gaps in financing and efficiency, and estimates of the need for additional spending on investment, operations, and maintenance. Each volume also comes with a detailed data appendix—providing easy access to all the relevant infrastructure indicators at the country level—which is a resource in and of itself.

In addition to these sector volumes, the AICD has produced a series of country reports that weave together all the findings relevant to one particular country to provide an integral picture of the infrastructure situation at the national level. Yet another set of reports provides an overall picture of the state of regional integration of infrastructure networks for each of the major regional economic communities of Sub-Saharan Africa. All of these papers are available through the project web portal, http://www.infrastructureafrica.org, or through the World Bank's Policy Research Working Paper series.

With the completion of this full range of analytical products, we hope to place the findings of the AICD effort at the fingertips of all interested policy makers, development partners, and infrastructure practitioners.

Vivien Foster and Cecilia Briceño-Garmendia

Acknowledgments

This book was co-authored by Anton Eberhard, Orvika Rosnes, Maria Shkaratan, and Haakon Vennemo, under the overall guidance of series editors Vivien Foster and Cecilia Briceño-Garmendia.

The book draws upon a number of background papers that were prepared by World Bank staff and consultants, under the auspices of the Africa Infrastructure Country Diagnostic (AICD). Key contributors to the book on a chapter-by-chapter basis were as follows.

Chapter 1

Contributors
Anton Eberhard, Vivien Foster, Cecilia Briceño-Garmendia, Maria Shkaratan, Fatimata Ouedraogo, Daniel Camos.

Key Source Document
Eberhard, Anton, Vivien Foster, Cecilia Briceño-Garmendia, Fatimata Ouedraogo, Daniel Camos, and Maria Shkaratan. 2008. "Underpowered: The State of the Power Sector in Sub-Saharan Africa." Background Paper 6, Africa Infrastructure Country Diagnostic, World Bank, Washington, DC.

Chapter 2

Contributors
Orvika Rosnes, Haakon Vennemo, Anton Eberhard, Vivien Foster, Cecilia Briceño-Garmendia, Maria Shkaratan, Fatimata Ouedraogo, Daniel Camos.

Key Source Documents
Rosnes, Orvika, and Haakon Vennemo. 2008. "Powering Up: Costing Power Infrastructure Spending Needs in Sub-Saharan Africa." Background Paper 5, Africa Infrastructure Country Diagnostic, World Bank, Washington, DC.

Eberhard, Anton, Vivien Foster, Cecilia Briceño-Garmendia, Fatimata Ouedraogo, Daniel Camos, and Maria Shkaratan. 2008. "Underpowered: The State of the Power Sector in Sub-Saharan Africa." Background Paper 6, Africa Infrastructure Country Diagnostic, World Bank, Washington, DC.

Chapter 3

Contributors
Orvika Rosnes, Haakon Vennemo, Anton Eberhard.

Key Source Document
Rosnes, Orvika, and Haakon Vennemo. 2008. "Powering Up: Costing Power Infrastructure Spending Needs in Sub-Saharan Africa." Background Paper 5, Africa Infrastructure Country Diagnostic, World Bank, Washington, DC.

Chapter 4

Contributors
Anton Eberhard, Vivien Foster, Cecilia Briceño-Garmendia, Maria Shkaratan, Maria Vagliasindi, John Nellis, Fatimata Ouedraogo, Daniel Camos.

Key Source Documents
Eberhard, Anton, Vivien Foster, Cecilia Briceño-Garmendia, Fatimata Ouedraogo, Daniel Camos, and Maria Shkaratan. 2008. "Underpowered: The State of the Power Sector in Sub-Saharan Africa." Background Paper 6, Africa Infrastructure Country Diagnostic, World Bank, Washington, DC.

Vagliasindi, Maria, and John Nellis. 2010. "Evaluating Africa's Experience with Institutional Reform for the Infrastructure Sectors." Working Paper 23, Africa Infrastructure Country Diagnostic, World Bank, Washington, DC.

Chapter 5

Contributors
Anton Eberhard, Vivien Foster, Cecilia Briceño-Garmendia, Sudeshna Ghosh Banerjee, Maria Shkaratan, Quentin Wodon, Amadou Diallo, Taras Pushak, Hellal Uddin, Clarence Tsimpo.

Key Source Document
Banerjee, Sudeshna, Quentin Wodon, Amadou Diallo, Taras Pushak, Hellal Uddin, Clarence Tsimpo, and Vivien Foster. 2008. "Access, Affordability and Alternatives: Modern Infrastructure Services in Sub-Saharan Africa." Background Paper 2, Africa Infrastructure Country Diagnostic, World Bank, Washington, DC.

Chapter 6

Contributors
Anton Eberhard, Vivien Foster, Cecilia Briceño-Garmendia, Maria Shkaratan, Fatimata Ouedraogo, Daniel Camos.

Key Source Documents
Eberhard, Anton, Vivien Foster, Cecilia Briceño-Garmendia, Fatimata Ouedraogo, Daniel Camos, and Maria Shkaratan. 2008. "Underpowered: The State of the Power Sector in Sub-Saharan Africa." Background Paper 6, Africa Infrastructure Country Diagnostic, World Bank, Washington, DC.
Briceño-Garmendia, Cecilia, and Maria Shkaratan. 2010. "Power Tariffs: Caught Between Cost Recovery and Affordability." Working Paper 8, Africa Infrastructure Country Diagnostic, World Bank, Washington, DC.

Chapter 7

Contributors
Maria Shkaratan, Cecilia Briceño-Garmendia, Karlis Smits, Vivien Foster, Nataliya Pushak, Jacqueline Irving, Astrid Manroth.

Key Source Documents
Briceño-Garmendia, Cecilia, Karlis Smits, and Vivien Foster. 2008. "Financing Public Infrastructure in Sub-Saharan Africa: Patterns and Emerging Issues." Background Paper 15, Africa Infrastructure Country Diagnostic, World Bank, Washington, DC.
Irving, Jacqueline, and Astrid Manroth. 2009. "Local Sources of Financing for Infrastructure in Africa: A Cross-Country Analysis." Policy Research Working Paper 4878, World Bank, Washington, DC.

Substantial sections of this book were derived from the above-listed background papers, as well as from the text of the AICD flagship report, *Africa's Infrastructure: A Time of Transformation*, edited by Vivien Foster and Cecilia Briceño-Garmendia.

The work benefited from widespread peer review from colleagues within the World Bank, notably Rob Mills, Dana Rysankova, Reto Thoenen, and Fabrice Karl Bertholet. The external peer reviewer for this volume, Mark Davis, provided constructive and thoughtful comments. The comprehensive editorial effort of Steven Kennedy is much appreciated.

Philippe Benoit, David Donaldson, Gabriel Goddard, S. Vijay Iyer, Luiz Maurer, Rob Mills, Fanny Missfeldt-Ringius, Lucio Monari, Kyran O'Sullivan, Prasad Tallapragada V.S.N., Clemencia Torres, and Tjaarda P. Storm van Leeuwen contributed significantly to the technical analysis and policy recommendations for the AICD power sector work, which formed the basis of this book.

None of this research would have been possible without the generous collaboration of government officials in the key sector institutions of each country, as well as the arduous work of local consultants who assembled this information in a standardized format. Key contributors to the book on a country-by-country basis were as follows.

Country	Local consultants or other partners
Angola	Fares Khoury (Etude Economique Conseil, Canada)
Benin	Jean-Marie Fansi (Pricewaterhouse Coopers)
Botswana	Adam Vickers, Nelson Mokgethi
Burkina Faso	Maxime Kabore
Cameroon	Kenneth Simo (Pricewaterhouse Coopers)
Cape Verde	Sandro de Brito
Central African Republic	Ibrahim Mamame
Chad	Kenneth Simo (Pricewaterhouse Coopers)
Congo, Dem. Rep.	Henri Kabeya
Congo, Rep.	Mantsie Rufin-Willy
Côte d'Ivoire	Jean-Phillipe Gogua, Roland Amehou
Ethiopia	Yemarshet Yemane
Gabon	Fares Khoury (Etude Economique Conseil, Canada)
Ghana	Afua Sarkodie
Kenya	Ayub Osman (Pricewaterhouse Coopers)
Lesotho	Peter Ramsden
Madagascar	Gerald Razafinjato
Mali	Ibrahim Mamame

Country	*Local consultants or other partners*
Mauritania	Fares Khoury (Etude Economique Conseil, Canada)
Mauritius	Boopen Seetanah
Mozambique	Manuel Ruas
Namibia	Peter Ramsden
Niger	Oumar Abdou Moulaye
Nigeria	Abiodun Momodu
Rwanda	Charles Uramutse
Sierra Leone	Adam Vickers, Nelson Mokgethi with the support of Alusine Kamara in SL Statistical Office
Senegal	Alioune Fall
South Africa	Peter Ramsden
Swaziland	Adam Vickers, Nelson Mokgethi
Tanzania	Adson Cheyo (Pricewaterhouse Coopers)
Uganda	Adson Cheyo (Pricewaterhouse Coopers)
Zambia	Mainza Milimo, Natasha Chansa (Pricewaterhouse Coopers)
Zimbabwe	Eliah Tafangombe

Abbreviations

All currency denominations are in U.S. dollars unless noted.

AICD	Africa Infrastructure Country Diagnostic
AIM	alternative investment market
AMADER	Agence Malienne pour le Developpement de l'Energie Domestique et d'Electrification Rurale
AU	African Union
BPC	Botswana Power Corporation
BRVM	Bourse Régionale des Valeurs Mobilières
capex	capital expenditures
CAPP	Central African Power Pool
CDM	Clean Development Mechanism
CER	certified emission reduction credit
CIE	Compagnie Ivoirienne d'Electricité
CIPREL	Compagnie Ivoirienne de Production d'Electricité
CREST	Commercial Reorientation of the Electricity Sector Toolkit
DBT	decreasing block tariff
EAPP	East African Power Pool
ECOWAS	Economic Community of West African States
EDF	Electricité de France

EDM	Electricidade de Moçambique
ESMAP	Energy Sector Management Assistance Program
FR	fixed rate
GDP	gross domestic product
GW	gigawatt
HFO	heavy fuel oil
IBT	increasing block tariff
ICT	information and communication technology
IDA	International Development Association
IFRS	International Financial Reporting Standards
IPP	independent power project
KenGen	Kenya Electricity Generating Company
KPLC	Kenya Power and Lighting Company
kVA	kilovolt-ampere
kWh	kilowatt-hour
LRMC	long-run marginal cost
LuSE	Lusaka Stock Exchange
MW	megawatt
NEPAD	New Partnership for Africa's Development
NES	National Electrification Scheme
NGO	nongovernmental organization
O&M	operations and maintenance
ODA	official development assistance
OECD	Organisation for Economic Co-operation and Development
opex	operational expenses
PPA	power-purchase agreement
PPI	private participation in infrastructure
PPIAF	Public-Private Infrastructure Advisory Facility
Q	quintile
REA	rural electrification agency
REF	rural electrification fund
RERA	Regional Electricity Regulators Association
ROR	rate of return
SADC	Southern African Development Community
SAPP	Southern African Power Pool
SHEP	Self-Help Electrification Programme
SOE	state-owned enterprise
SSA	Sub-Saharan Africa
T&D	transmission and distribution

tcf	trillion cubic feet
TFP	total factor productivity
TOU	time of use
TPA	third-party access
TW	terawatt
TWh	terawatt-hour
UCLF	unplanned capability loss factor
USO	universal service obligation
WAPP	West African Power Pool
Wh	watt-hour
WSS	water supply and sanitation

CHAPTER 1

Africa Unplugged

Sub-Saharan Africa is in the midst of a power crisis. The region's power generation capacity is lower than that of any other world region, and capacity growth has stagnated compared with other developing regions. Household connections to the power grid are scarcer in Sub-Saharan Africa than in any other developing region.

The average price of power in Sub-Saharan Africa is double that in other developing regions, but the supply of electrical power is unreliable throughout the continent. The situation is so dire that countries increasingly rely on emergency power to cope with electricity shortages.[1] The weakness of the power sector has constrained economic growth and development in the region.

The Region's Underdeveloped Energy Resources

An estimated 93 percent of Africa's economically viable hydropower potential—or 937 terawatt-hours (TWh) per year, about one-tenth of the world's total—remains unexploited. Much of that is located in the Democratic Republic of Congo, Ethiopia, Cameroon, Angola, Madagascar, Gabon, Mozambique, and Nigeria (in descending order by capacity). Some of the largest operating hydropower installations are in

1

the Democratic Republic of Congo, Mozambique, Nigeria, Zambia, and Ghana. Burundi, Lesotho, Malawi, Rwanda, and Uganda also rely heavily on hydroelectricity.

Although most Sub-Saharan African countries have some thermal power stations, only a few use local petroleum and gas resources. Instead, most countries rely on imports. There are a few exceptions: proven oil reserves are concentrated in Nigeria (36 billion barrels), Angola (9 billion barrels), and Sudan (6.4 billion barrels). A number of smaller deposits have been found in Gabon, the Republic of Congo, Chad, Equatorial Guinea, Cameroon, the Democratic Republic of Congo, and Côte d'Ivoire.[2] Overall, Sub-Saharan Africa accounts for less than 5 percent of global oil reserves. Actual oil production follows a similar pattern (BP 2007).

Natural gas reserves are concentrated primarily in Nigeria (5.2 trillion cubic feet [tcf]). Significant natural gas discoveries have also been made in Mozambique, Namibia, and Angola, with reserves of 4.5 tcf, 2.2 tcf, and 2.0 tcf, respectively. Small amounts have been discovered in Tanzania. Gas reserves in Sub-Saharan Africa make up less than 4 percent of the world's total proven reserves, and actual gas production is an even smaller proportion of the world's total production (BP 2007).

Only one nuclear power plant has been built on the continent: the 1,800 megawatt (MW) Koeberg station in South Africa. Africa's natural uranium reserves account for approximately one-fifth of the world's total and are located mainly in South Africa, Namibia, and Niger.

Geothermal power looks economically attractive in the Rift Valley, and Kenya has several geothermal plants in operation. The continent has abundant renewable energy resources, particularly solar and wind, although these are often costly to develop and mostly provide off-grid power in remote areas where alternatives such as diesel generators are expensive.

The Lag in Installed Generation Capacity

The combined power generation capacity of the 48 countries of Sub-Saharan Africa is 68 gigawatts (GW)—no more than that of Spain. Excluding South Africa, the total falls to 28 GW, equivalent to the installed capacity of Argentina (data for 2005; EIA 2007). Moreover, as much as 25 percent of installed capacity is not operational for various reasons, including aging plants and lack of maintenance.

The installed capacity per capita in Sub-Saharan Africa (excluding South Africa) is a little more than one-third of South Asia's (the two regions were equal in 1980) and about one-tenth of that of Latin America (figure 1.1). Capacity growth has been largely stagnant during the past three decades, with growth rates of barely half those found in other developing regions. This has widened the gap between Sub-Saharan Africa and the rest of the developing world, even compared with other country groups in the same income bracket (Yepes, Pierce, and Foster 2008).

South Africa's power infrastructure stands in stark contrast to that of the region as a whole. With a population of 47 million people, South Africa has a total generation capacity of about 40,000 MW. Nigeria comes in second, with less than 4,000 MW, despite its much larger population of 140 million. A handful of countries have intermediate capacity: the Democratic Republic of Congo (2,443 MW), Zimbabwe (2,099 MW), Zambia (1,778 MW), Ghana (1,490 MW), Kenya (1,211 MW), and Côte d'Ivoire (1,084 MW)—although not all of their capacity is operational. Capacity is much lower in other countries: Mali (280 MW), Burkina Faso (180 MW), Rwanda (31 MW), and Togo (21 MW) (EIA 2007). Per capita generation capacity also varies widely among countries (figure 1.2).

In 2004, the power plants of Sub-Saharan Africa generated 339 TWh of electricity—approximately 2 percent of the world's total. South African power plants generated about 71 percent of that total (Eberhard and others 2008). Coal-fired plants generate 93 percent of South Africa's electricity, and coal is therefore the dominant fuel in the region. Most of

Figure 1.1   Power Generation Capacity by Region, 1980–2005

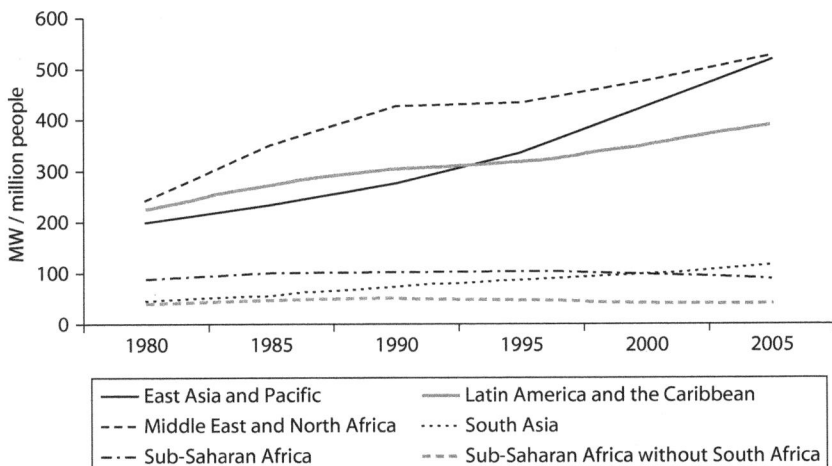

Source: Derived by authors from AICD 2008 and EIA 2007.
Note: MW = megawatt.

Figure 1.2   Power Generation Capacity in Sub-Saharan Africa by Country, 2006

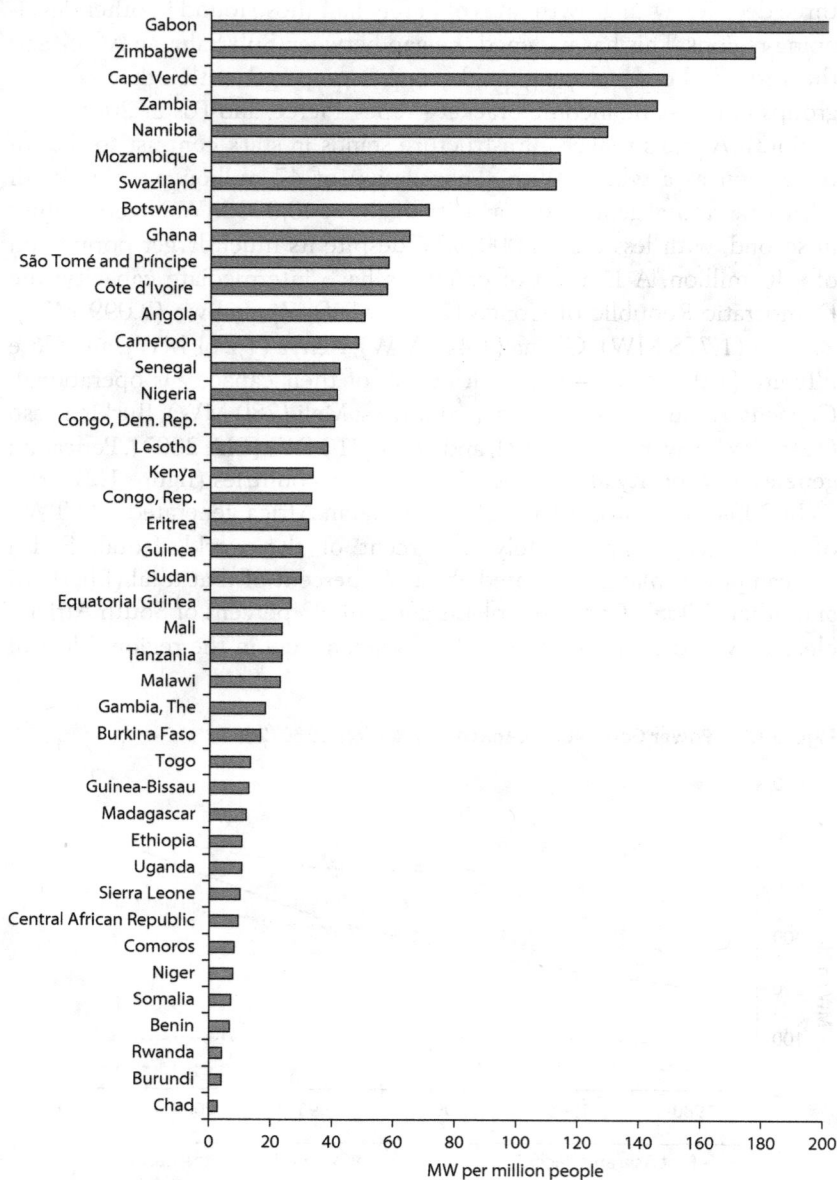

Source: EIA 2007.
Note: By comparison, South Africa's figure is 855 MW per million people. MW = megawatt.

the region's coal reserves are located in the south, mainly in South Africa, which has the fifth-largest reserves globally and ranks fifth in annual global production (BP 2007). Few other countries in the region rely on coal, but Botswana and Zimbabwe are among the exceptions.[3] Coal reserves in Africa constitute just 5.6 percent of the global total.

Power generation in Sub-Saharan Africa is much different outside of South Africa. Hydropower accounts for close to 70 percent of electricity generation (and about 50 percent of installed generation capacity), with the remainder divided almost evenly between oil and natural gas generators.

Stagnant and Inequitable Access to Electricity Services

Sub-Saharan Africa has low rates of electrification. Less than 30 percent of the population of Sub-Saharan Africa has access to electricity, compared with about 65 percent in South Asia and more than 90 percent in East Asia (figure 1.3). Based on current trends, fewer than 40 percent of

Figure 1.3   Household Electrification Rate in World Regions, 1990–2005

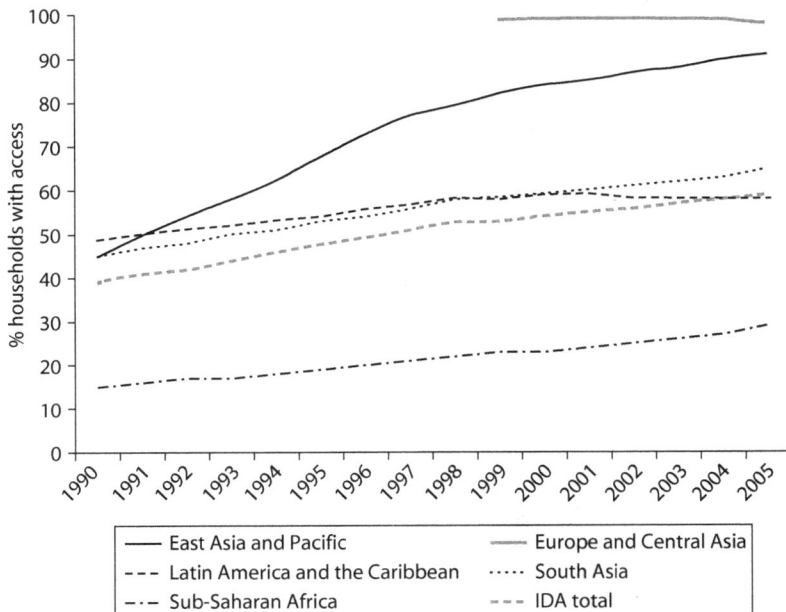

	East Asia and Pacific		Europe and Central Asia
	Latin America and the Caribbean		South Asia
	Sub-Saharan Africa		IDA total

Source: Eberhard and others 2008.
Note: IDA = International Development Association.

African countries will achieve universal access to electricity by 2050 (Banerjee and others 2008).

Per capita consumption of electricity averages just 457 kilowatt-hour (kWh) annually in the region, and that figure falls to 124 kWh if South Africa is excluded (Eberhard and others 2008). By contrast, the annual average per capita consumption in the developing world is 1,155 KWh and 10,198 kWh. If South Africa is excluded, Sub-Saharan Africa is the only world region in which per capita consumption of electricity is falling.

Figure 1.4 shows the relationship between electricity consumption and economic development in world regions. All countries in Sub-Saharan Africa (except South Africa) lag far behind other regions in per capita power consumption and gross domestic product (GDP).

Because of its low electricity consumption, Sub-Saharan Africa is an insignificant contributor to carbon dioxide emissions and climate change. It has the lowest per capita emissions among all world regions and has some of the lowest emissions in terms of GDP output. Excluding South Africa, the power sector in Sub-Saharan Africa accounts for less than 1 percent of global carbon dioxide emissions.

Figure 1.4   Per Capita Electricity Consumption and GDP in Selected Countries of Sub-Saharan Africa and World Regions, 2004

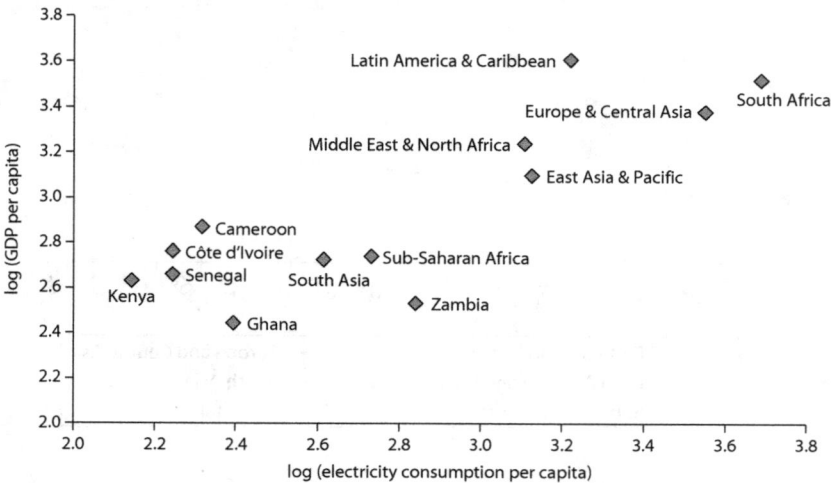

Source: Eberhard and others 2008.
Note: GDP = gross domestic product.

Unreliable Electricity Supply

Power supply in Sub-Saharan Africa is notoriously unreliable. Conventional measures of the reliability of power systems include the unplanned capability loss factor (UCLF)[4] of generators, the number of transmission interruptions, and indexes of the frequency and duration of interruptions. Yet most African countries still do not systematically collect or report these data. The World Bank enterprise surveys, which provide a useful alternative measure of the reliability of grid-supplied power, indicate that most African enterprises experience frequent outages. In 2007, for example, firms in Senegal, Tanzania, and Burundi experienced power outages for an average of 45, 63, and 144 days, respectively (figure 1.5).

The Prevalence of Backup Generators

In countries that report more than 60 days of power outages per year, firms identify power as a major constraint to doing business and are more likely to own backup generators. The size, sector, and export orientation of the firm also influence the likelihood of the firm having its own generation facilities (hereafter own generation). Larger firms are more likely to own backup generators (figure 1.6).

Own generation constitutes a significant proportion of total installed power capacity in the region—as much as 19 percent in West Africa (figure 1.7). In the Democratic Republic of Congo, Equatorial Guinea, and Mauritania, backup generators account for half of total installed capacity. The share is much lower in southern Africa, but it is likely to increase as the region experiences further power outages. South Africa—which for many years maintained surplus capacity—recently experienced acute power shortages. The value of in-house generating capacity in Sub-Saharan Africa as a percentage of gross fixed capital formation ranges from 2 percent to as high as 35 percent (Foster and Steinbuks 2008).

Frequent power outages result in forgone sales and damaged equipment for businesses, which result in significant losses. These losses are equivalent to 6 percent of turnover on average for firms in the formal sector and as much as 16 percent of turnover for informal sector enterprises that lack a backup generator (Foster and Steinbuks 2008).

The overall economic costs of power outages are substantial. Based on outage data from the World Bank's Investment Climate Assessments (ICA), utility load-shedding data,[5] and the estimates of the value of lost load or unserved energy, power outages in the countries in Sub-Saharan Africa constitute an average of 2.1 percent of GDP. In those Africa

Figure 1.5   Power Outages, Days per Year, 2007–08

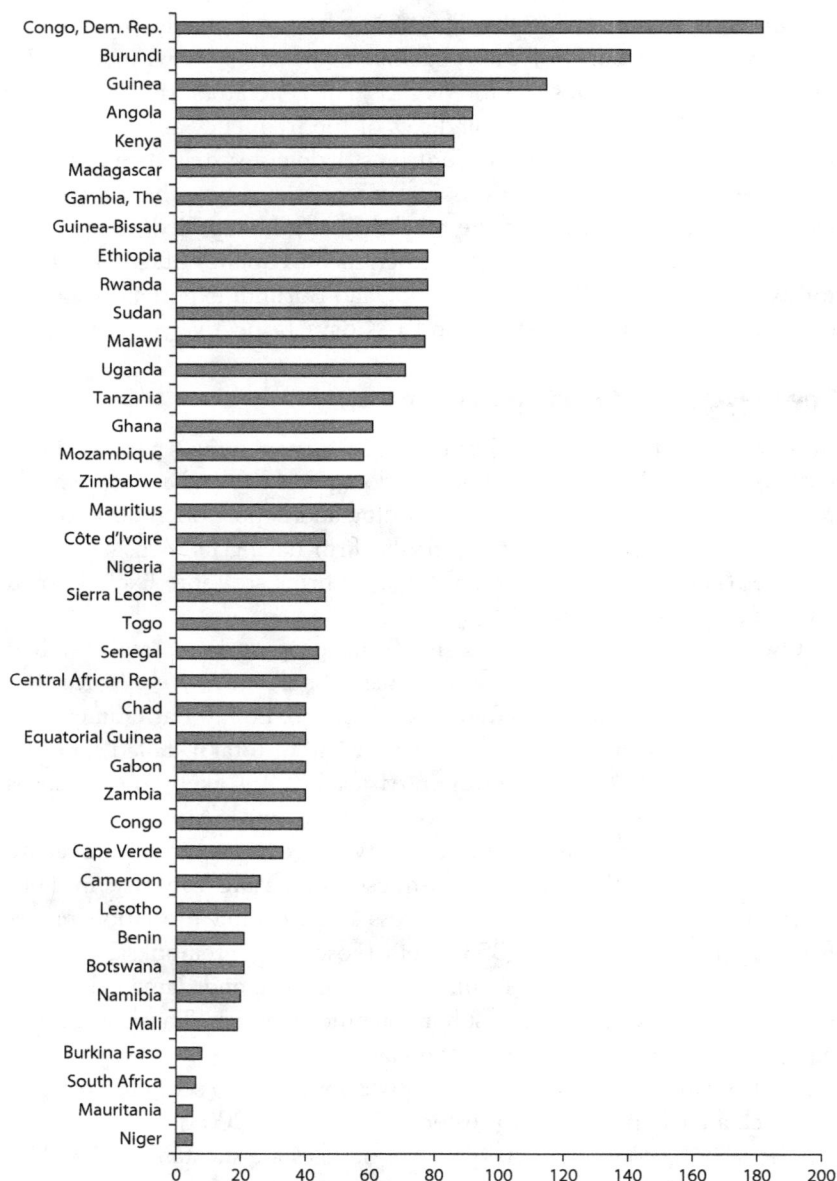

Source: Enterprise Survey database; World Bank 2008.

Figure 1.6   Generator Ownership by Firm Size

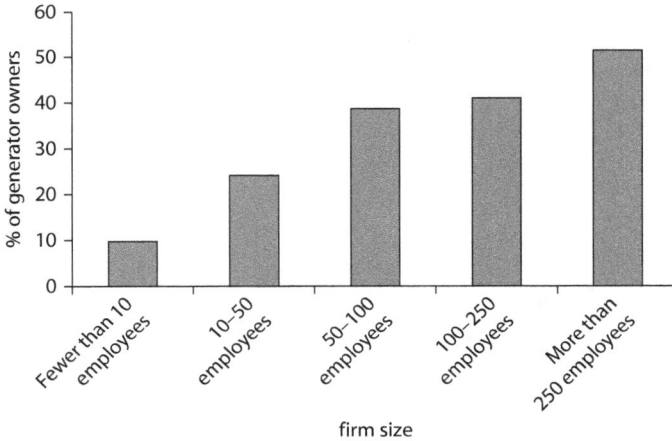

Source: Foster and Steinbuks 2008.

Figure 1.7   Own Generation as Share of Total Installed Capacity by Subregion, 2006

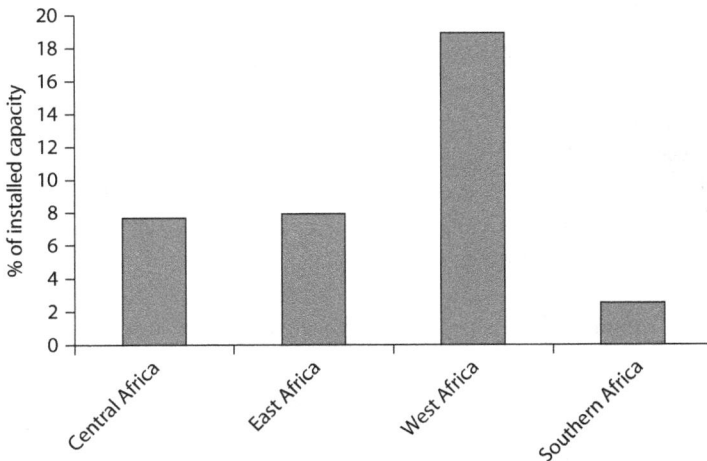

Source: Foster and Steinbuks 2008.

Infrastructure Country Diagnostic (AICD) countries for which we were able to make our own calculations (about half of the countries), the costs ranged from less than 1 percent of GDP in countries such as Niger to 4 percent of GDP and higher in countries such as Tanzania (figure 1.8).

Figure 1.8   Economic Cost of Power Outages as Share of GDP, 2005

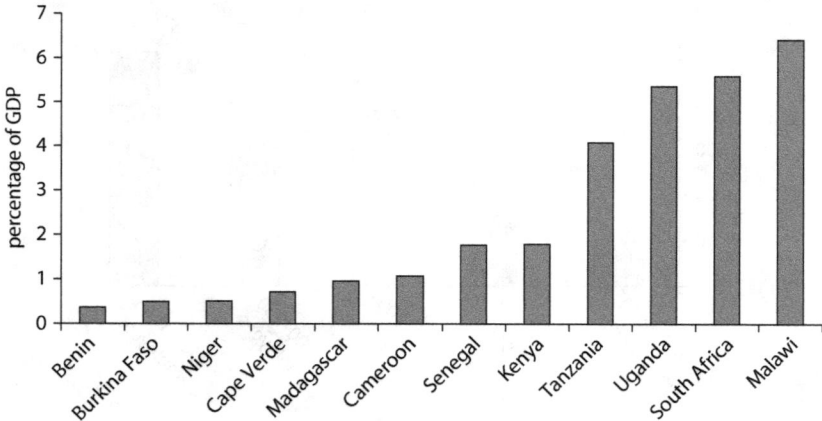

Source: Briceño-Garmendia 2008 and authors' calculations of own-generation costs based on Foster and Steinbuks 2008.
Note: GDP = gross domestic product.

Increasing Use of Leased Emergency Power

The increasing use of grid-connected emergency power in the region reflects the gravity of the power crisis (table 1.1). Countries experiencing pressing power shortages can enter into short-term leases with specialized operators who install new capacity (typically in shipping containers) within a few weeks, which is much faster than a traditional power-generation project. The country leases the equipment for a few months to a few years, after which the private operator removes the power plant. Temporary emergency generators now account for an estimated 750 MW of capacity in Sub-Saharan Africa, and they constitute a significant proportion of total capacity in some countries. Emergency power is relatively expensive—typically around $0.20–0.30 per kWh. In some countries, the cost of emergency power is a considerable percentage of GDP.[6] Procurement has also been tainted by corruption and bribery. For example, the Tanzanian prime minister and energy minister resigned in February 2008 after a parliamentary investigation revealed that lucrative contracts for emergency power had been placed with a company with no power generation experience.

Despite the high cost of leased power, a multi-megawatt emergency power installation can be large enough to achieve economies of scale, and it is a better option than individual backup generators. The cost of

Table 1.1 Overview of Emergency Power Generation in Sub-Saharan Africa (Up to 2007)

Country	Date	Contract duration (years)	Emergency capacity (MW)	Percent of total installed capacity	Estimated annual cost (% GDP)
Angola	2006	2	150	18.1	1.04
Gabon	—	—	14	3.4	0.45
Ghana	2007	1	80	5.4	1.90
Kenya	2006	1	100	8.3	1.45
Madagascar	2004	Several years	50	35.7	2.79
Rwanda	2005	2	15	48.4	1.84
Senegal	2005	2	40	16.5	1.37
Sierra Leone	2007	1	20	100	4.25
Tanzania	2006	2	180	20.4	0.96
Uganda	2006	2	100	41.7	3.29

Source: Eberhard and others 2008.

Note: Leases for emergency power are generally short term. Therefore, installed capacities in individual countries change from year to year. — = Not available.

emergency power also far exceeds the value of lost load. Countries that have entered into these expensive, short-term contracts understand the potentially greater economic cost of power shortages.

A Power Crisis Exacerbated by Drought, Conflict, and High Oil Prices

In recent years, external factors have exacerbated the already precarious power situation in Sub-Saharan Africa. Drought has seriously reduced the power available to hydro-dependent countries in western and eastern Africa. Countries with significant hydropower installations in affected catchments—Burundi, Ghana, Kenya, Madagascar, Rwanda, Tanzania, and Uganda—have had to switch to expensive diesel power. High international oil prices have also put enormous pressure on all of the oil-importing countries of Sub-Saharan Africa, especially those dependent on diesel and heavy fuel oil for their power-generation needs. Furthermore, war has seriously damaged power infrastructure in the Central African Republic, the Democratic Republic of Congo, Liberia, Sierra Leone, and Somalia. In Zimbabwe, political conflict and economic contraction have undermined the power system as investment resources have dried up. Overall, countries in conflict perform worse in the development of infrastructure than do countries at peace (Yepes, Pierce, and Foster 2008). Other countries, such as Nigeria and South Africa, are experiencing a power crisis induced by rapid growth in electricity demand coupled with prolonged underinvestment in new generation capacity. Both of those countries have experienced blackouts in recent years.

High Power Prices That Generally Do Not Cover Costs

Power in Sub-Saharan Africa is generally expensive by international standards (figure 1.9). The average power tariff in Sub-Saharan Africa is $0.12 per kWh, which is about twice the tariff in other parts of the developing world, and almost as high as in the high-income countries of the Organisation for Economic Co-operation and Development. There are exceptions: Angola, Malawi, South Africa, Zambia, and Zimbabwe have maintained low prices that are well below costs (Sadelec 2006).

Power from backup generators is much more expensive than grid power (figure 1.10), which increases the weighted average cost of power to consumers above the figures quoted previously.

Figure 1.9   Average Residential Electricity Prices in Sub-Saharan Africa and Other Regions, 2005

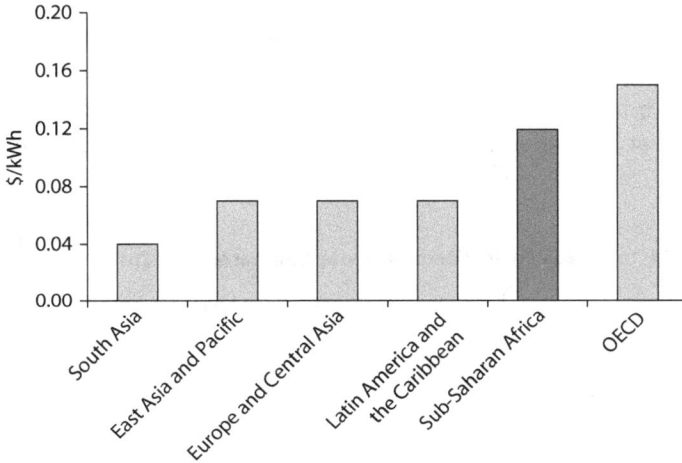

Source: Briceño-Garmendia and Shkaratan 2010.
Note: OECD = Organisation for Economic Co-operation and Development.

Figure 1.10   Average Cost of Grid and Backup Power in Sub-Saharan Africa

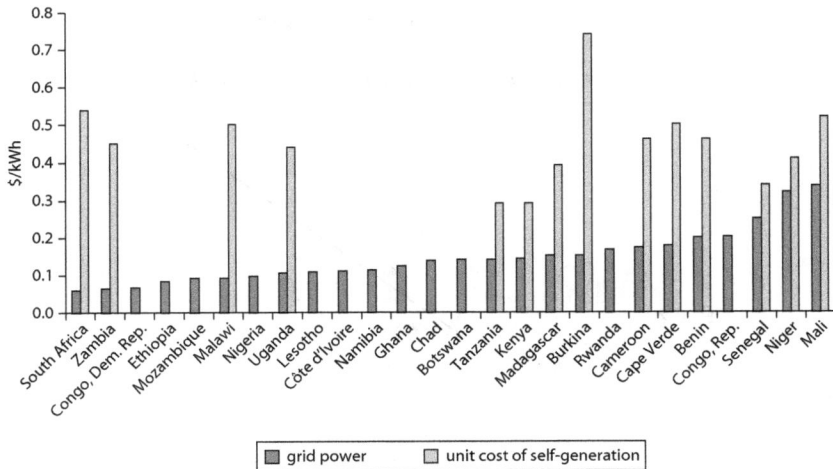

Source: Briceño-Garmendia 2008 and authors' calculations of own-generation costs based on Foster and Steinbuks 2008.

Although electricity in the region is relatively expensive, most Sub-Saharan Africa countries are doing little more than covering their average operating costs (figure 1.11). The close correlation between average effective tariff[7] and average cost across the countries of Sub-Saharan Africa (as high as 58 percent) indicates that for the most part they price their power with the intent of breaking even. Countries with average operating costs in excess of $0.15 per kWh tend to set prices somewhat below this level.

Figure 1.11   Average Power Sector Revenue Compared with Costs

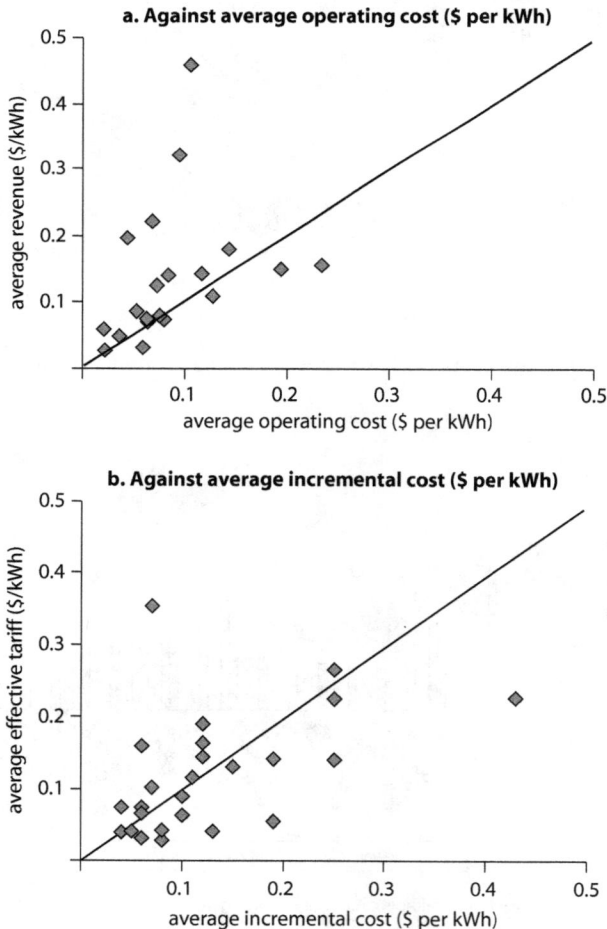

a. Against average operating cost ($ per kWh)

b. Against average incremental cost ($ per kWh)

Source: Briceño-Garmendia and Shkaratan 2010.

A simple comparison of average revenues and average operating costs misrepresents the prospects for long-term cost recovery for two reasons. First, owing to major failures in utility revenue collection, operators collect far less per unit of electricity from customers than they charge. Second, for many countries in Sub-Saharan Africa, the average total cost associated with power developments in the *past* is actually higher than the average incremental cost of producing new power in the *future*. This is because historically, power development has been done using small-scale and inefficient generation technologies, which could be superseded as countries become able to trade power with one another, thereby harnessing larger-scale and more efficient forms of production. Thus, a comparison of the average tariff that operators charge (but do not necessarily collect) with the average incremental cost of generating power provides a more accurate picture of the situation. Regardless, in some countries, revenues would cover costs only if tariffs were fully collected and if the power system moved toward a more efficient production structure.

In the past the state or donors have subsidized the share of capital investment that tariffs could not cover.[8] Households account for the majority of power utility sales in many African countries but only about 50 percent of sales revenue because of poor collections and underpricing. Thus, tariffs charged to commercial and industrial consumers are important sources of revenue for the utility. It is more difficult to assess whether tariffs for commercial and industrial customers are high enough to cover costs. The limited evidence available suggests that the average revenue raised from low- and medium-voltage customers does cover costs, whereas high-voltage customers tend to pay less. This relative price differential, which is not uncommon around the world, reflects the fact that high-voltage customers take their supply directly from the transmission grid. They do not make use of the distribution network and hence do not create such high costs for the power utility. Nevertheless, it is unclear whether these lower tariffs for large, high-voltage customers are actually covering costs.

Numerous countries have historically charged highly discounted tariffs of just a few cents per kWh to large-scale industrial and mining customers, such as the aluminum smelting industry in Cameroon, Ghana, and South Africa and the mining industry in Zambia. These arrangements were intended to secure base-load demand to support the development of large-scale power projects that went beyond the immediate demands of the country. Growing demand has begun to absorb excess capacity, however, which makes the relevance of the discounts dubious.

Deficient Power Infrastructure Constrains Social and Economic Development

Based on panel data analysis, Calderón (2008) provides a comprehensive assessment of the impact of infrastructure stocks on growth in Sub-Saharan Africa between the early 1990s and the early 2000s. Calderón finds that if African countries were to catch up with the regional leader, Mauritius, in terms of infrastructure stock and quality, their per capita economic growth rates would increase by an average of 2.2 percent per year. Catching up with the East Asian median country, the Republic of Korea, would bring gains of 2.6 percent per year. In several countries— including Côte d'Ivoire, the Democratic Republic of Congo, and Senegal—the effect would be even greater.

Deficient power infrastructure and power outages dampen economic growth, especially through their detrimental effect on firm productivity. Using enterprise survey data collected through the World Bank's Investment Climate Assessments, Escribano, Guasch, and Peña (2008) find that in most countries of Sub-Saharan Africa, infrastructure accounts for 30–60 percent of the effect of investment climate on firm productivity—well ahead of most other factors, including red tape and corruption. In half of the countries analyzed, the power sector accounted for 40–80 percent of the infrastructure effect (figure 1.12).

Infrastructure is also an important input into human development. Better provision of electricity improves health care because vaccines and medications can be safely stored in hospitals and food can be preserved at home (Jimenez and Olson 1998). Electricity also improves literacy and primary school completion rates because students can read and study when there is no natural light (Barnes 1988; Brodman 1982; Foley 1990; Venkataraman 1990). Similarly, better access to electricity lowers costs for businesses and increases investment, driving economic growth (Reinikka and Svensson 1999).

In summary, chronic power problems—including insufficient investment in generation capacity and networks, stagnant or declining connectivity, poor reliability, and high costs and prices (which further hinders maintenance, refurbishment, and system expansion)—have created a power crisis in Sub-Saharan Africa. Drought, conflict, and high oil prices have exacerbated the crisis. The overall deficiency of the power sector has constrained economic and social development. Although the extent of the problems and challenges differs across regions and countries, Sub-Saharan Africa has generally lagged behind other regions of the world in terms of infrastructure and power sector investment and performance. This book investigates how these problems and challenges might be addressed.

Figure 1.12   Contribution of Infrastructure to Total Factor Productivity (TFP) of Firms

a. Overall contribution of infrastructure

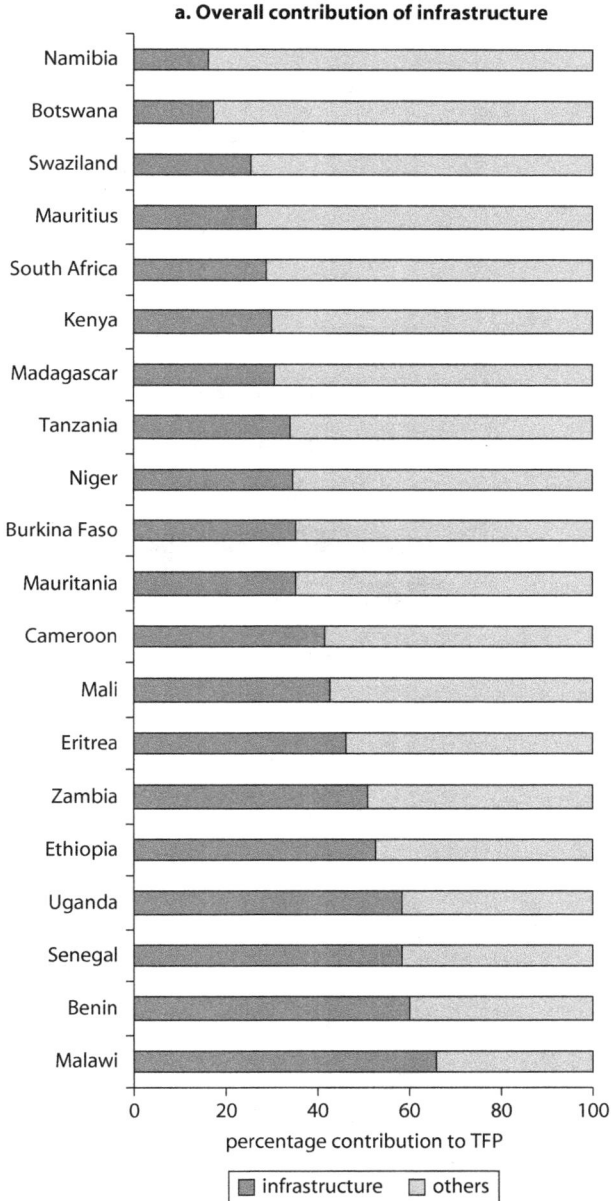

percentage contribution to TFP

■ infrastructure   □ others

(continued next page)

Figure 1.12   *(continued)*

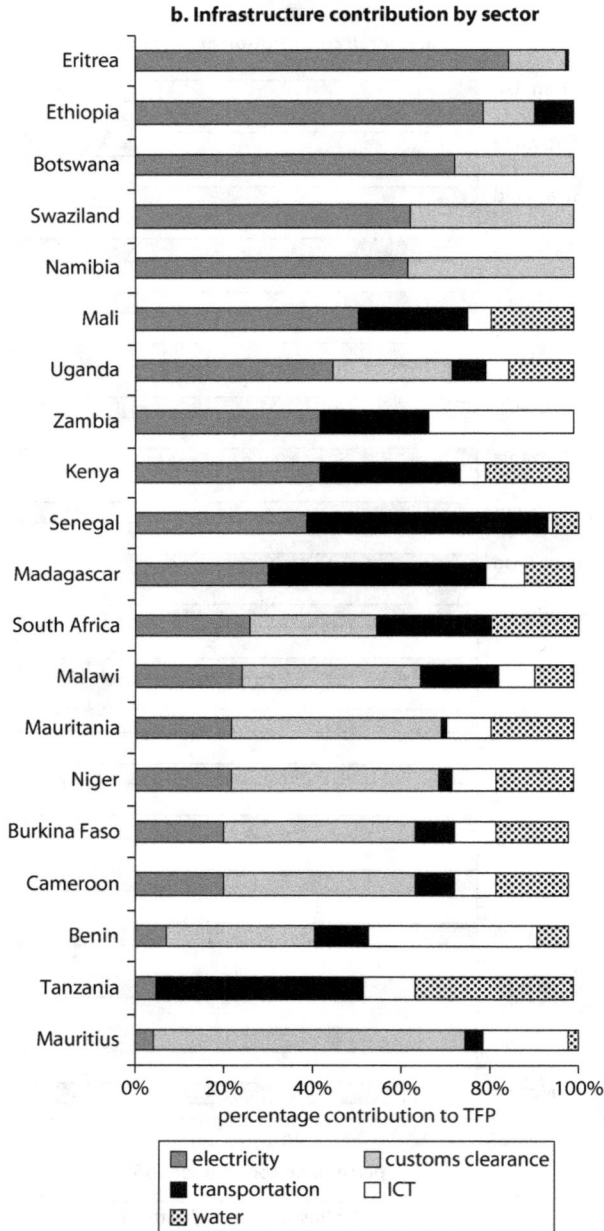

b. Infrastructure contribution by sector

Source: Escribano, Guasch, and Peña 2008.
Note: ICT = information and communication technology.

Notes

1. Emergency power is a term for expensive, short-term leases for generation capacity.
2. Small deposits were also recently discovered in countries such as Ghana and Uganda.
3. Mauritius, Namibia, Niger, and Tanzania also have small coal-generation plants. Mozambique is planning investments in coal power stations.
4. The UCLF is the percentage of time over a year that the generation plant is not producing power, excluding the time that the plant was shut down for routine, planned maintenance.
5. Load shedding occurs when the power grid is unable to meet demand, and customers' supply is cut off.
6. Spending on emergency power can displace expenditures on social services such as health and education. For example, Sierra Leone has a population of 6 million but only 28,000 electricity customers. The country relies heavily on an overpriced emergency diesel-based power supply contract for its electricity needs. As a result, the government of Sierra Leone has not been able to meet the minimum targets for expenditures in health and education that are required for continued budget support by the European Union and other donors.
7. Effective tariffs are prices per kWh at typical monthly consumption levels calculated using tariff schedules applicable to typical customers within each customer group.
8. One of the casualties of insufficient revenue is maintenance expenditure. Utility managers often have to choose between paying salaries, buying fuel, or purchasing spares (often resorting to cannibalizing parts from functional equipment). For example, in Sierra Leone, the overhead distribution network for the low-income eastern part of Freetown has been cannibalized for spare parts to repair the network of the high-income western part of the town. Thus, even with the advent of emergency generators, many former customers in the eastern districts remain without power.

References

AICD (Africa Infrastructure Country Diagnostic). 2008. *AICD Power Sector Database*. Washington, DC: World Bank.

Banerjee, Sudeshna, Quentin Wodon, Amadou Diallo, Taras Pushak, Hellal Uddin, Clarence Tsimpo, and Vivien Foster. 2008. "Access, Affordability and Alternatives: Modern Infrastructure Services in Sub-Saharan Africa." Background Paper 2, Africa Infrastructure Country Diagnostic, World Bank, Washington, DC.

Barnes, Douglas F. 1988. *Electric Power for Rural Growth: How Electricity Affects Rural Life in Developing Countries.* Boulder: Westview Press.

BP (British Petroleum). 2007. *Statistical Review of Energy.* London: Beacon Press.

Briceño-Garmendia, Cecilia. 2008. "Quasi-Fiscal Costs: A Never Ending Concern." Internal Note, World Bank, Washington, DC.

Briceño-Garmendia, Cecilia, and Maria Shkaratan. 2010. "Power Tariffs: Caught between Cost Recovery and Affordability." Working Paper 20, Africa Infrastructure Country Diagnostic, World Bank, Washington, DC.

Brodman, Janice. 1982. "Rural Electrification and the Commercial Sector in Indonesia." Discussion Paper D-73L, Resources for the Future, Washington, DC.

Calderón, Cesar. 2008. "Infrastructure and Growth in Africa." Working Paper 3, Africa Infrastructure Country Diagnostic, World Bank, Washington, DC.

Eberhard, Anton, Vivien Foster, Cecilia Briceño-Garmendia, Fatimata Ouedraogo, Daniel Camos, and Maria Shkaratan. 2008. "Underpowered: The State of the Power Sector in Sub-Saharan Africa." Background Paper 6, Africa Infrastructure Country Diagnostic, World Bank, Washington, DC.

EIA (Energy Information Administration). 2007. "International Energy Data." U.S. Department of Energy. http://www.eia.doe.gov/emeu/international.

Escribano, Alvaro, J. Luis Guasch, and Jorge Peña. 2008. "Impact of Infrastructure Constraints on Firm Productivity in Africa." Working Paper 9, Africa Infrastructure Country Diagnostic, World Bank, Washington, DC.

Foley, Gerald. 1990. *Electricity for Rural People.* London: Panos Institute.

Foster, Vivien, and Jevgenijs Steinbuks. 2008. "Paying the Price for Unreliable Power Supplies: In-House Generation of Electricity by Firms in Africa." Working Paper 2, Africa Infrastructure Country Diagnostic, World Bank, Washington, DC.

Jimenez, Antonio, and Ken Olson. 1998. "Renewable Energy for Rural Health Clinics." National Renewable Energy Laboratory, Golden, CO. http://www.nrel.gov/docs/legosti/fy98/25233.pdf.

Reinikka, Ritva, and Jakob Svensson. 1999. "Confronting Competition: Firms' Investment Response and Constraints in Uganda." In *Assessing an African Success: Farms, Firms, and Government in Uganda's Recovery,* ed. P. Collier and R. Reinikka, 207–34. Washington, DC: World Bank.

Sadelec, Ltd. 2006. "Electricity Prices in Southern and East Africa (Including Selected Performance Indicators)." Sadelec, Ltd., Johannesburg, South Africa.

Venkataraman, Krishnaswami. 1990. "Rural Electrification in the Asian and Pacific Region." In *Power Systems in Asia and the Pacific, with Emphasis on Rural*

Electrification, ed. Economic and Social Commission for Asia and the Pacific, 310–32. New York: United Nations.

———. 2008. *Enterprise Survey Database.* Washington, DC: World Bank.

Yepes, Tito, Justin Pierce, and Vivien Foster. 2008. "Making Sense of Africa's Infrastructure Endowment: A Benchmarking Approach." Policy Research Working Paper 4912, World Bank, Washington, DC.

CHAPTER 2

The Promise of Regional Power Trade

Africa consists of many small isolated economies. Integrating physical infrastructure is therefore necessary to promote regional economic integration and enable industries to reach economies of scale. In particular, regional integration would allow countries to form regional power pools, which can already be found at varying stages of maturity in Southern, West, East, and Central Africa. Regional trade would allow countries to substitute hydropower for thermal power, which would lead to a substantial reduction in operating costs—despite the requisite investments in infrastructure and cross-border transmission capacity. Our modeling indicates that the annual costs of power system operation and development in the region could fall by $2.7 billion. The returns to cross-border transmission investment could be 20–30 percent in most power pools and can be as high as 120 percent in the Southern African Power Pool (SAPP). The greater share of hydropower associated with regional trade would also reduce annual carbon dioxide emissions by 70 million tons.

Under regional power trade, a few large exporting countries would serve many power importers. The Democratic Republic of Congo, Ethiopia, and Guinea would emerge as the major hydropower exporters.

Yet the magnitude of the investments needed to develop their exporting potential is daunting relative to the size of their economies. At the same time, as many as 16 African countries would benefit (from a purely economic standpoint) from the opportunity to reduce costs by importing more than 50 percent of their power. Savings for those countries range from $0.01 to $0.07 per kilowatt-hour (kWh). The largest beneficiaries of regional trade would be smaller nations that lack domestic hydropower resources. For these countries, the cost savings generated by regional trade would repay the requisite investment in cross-border transmission in less than a year, contingent on neighboring countries developing sufficient surplus power to export.

Uneven Distribution and Poor Economies of Scale

Only a small fraction of the ample hydropower and thermal energy resources in Sub-Saharan Africa have been developed into power generation capacity. Some of the region's least expensive sources of power are far from major centers of demand in countries too poor to develop them. For example, 61 percent of regional hydropower potential is found in just two countries: the Democratic Republic of Congo and Ethiopia. Both are poor countries with a gross domestic product (GDP) of less than $30 billion.

The uneven distribution of resources in the region has forced many countries to adopt technically inefficient forms of generation powered by expensive imported fuels to serve their small domestic power markets. Expensive diesel or heavy fuel oil generators account for about one-third of installed capacity in Eastern and Western Africa (figure 2.1a). In many cases, countries that lack adequate domestic energy resources could replace this capacity with the much cheaper hydro and gas resources of neighboring countries.

Few countries in the region have sufficient demand to justify power plants large enough to exploit economies of scale (figure 2.1b). For example, 33 out of 48 countries in Sub-Saharan Africa have national power systems that produce and consume less than 500 megawatts (MW), and 11 countries have national power systems of less than 100 MW. The small market size of most countries in Sub-Saharan Africa contributes to severely inflated generation costs.

A comparison of operating costs disaggregated into four categories reveals the negative consequences of technically inefficient power

Figure 2.1   Profile of Power Generation Capacity in Sub-Saharan Africa

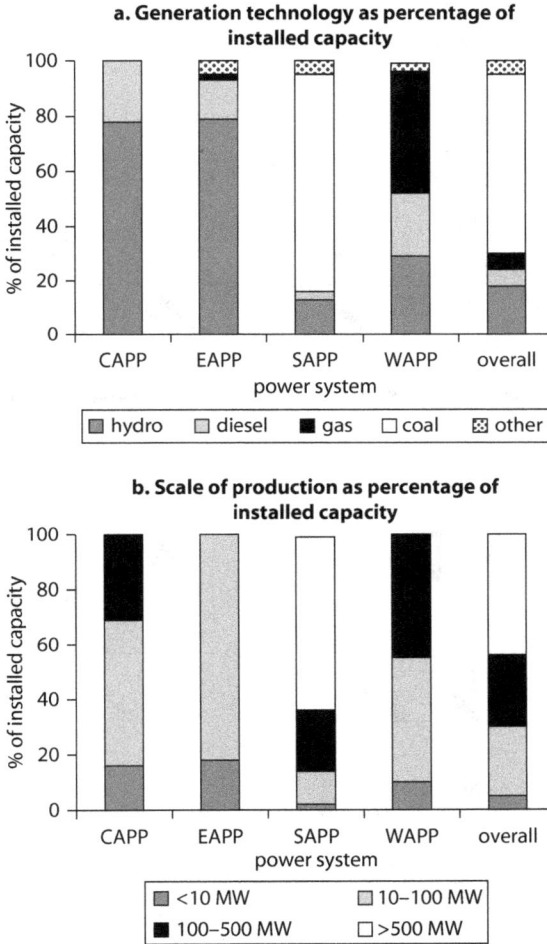

a. Generation technology as percentage of installed capacity

hydro   diesel   gas   coal   other

b. Scale of production as percentage of installed capacity

<10 MW   10–100 MW
100–500 MW   >500 MW

Source: Eberhard and others 2008.
Note: CAPP = Central African Power Pool; EAPP = East African Power Pool; SAPP = Southern African Power Pool; WAPP = West African Power Pool; MW = megawatt.

generation (figure 2.2). For example, the average operating cost of predominantly diesel-based power systems can be as high as $0.14 per kWh—almost twice the cost of predominantly hydro-based systems. Similarly, operating costs in countries with small national power systems (less than 200 MW installed capacity) are much higher than in countries with large national power systems (more than 500 MW

Figure 2.2   Disaggregated Operating Costs for Power Systems in Sub-Saharan Africa, 2005

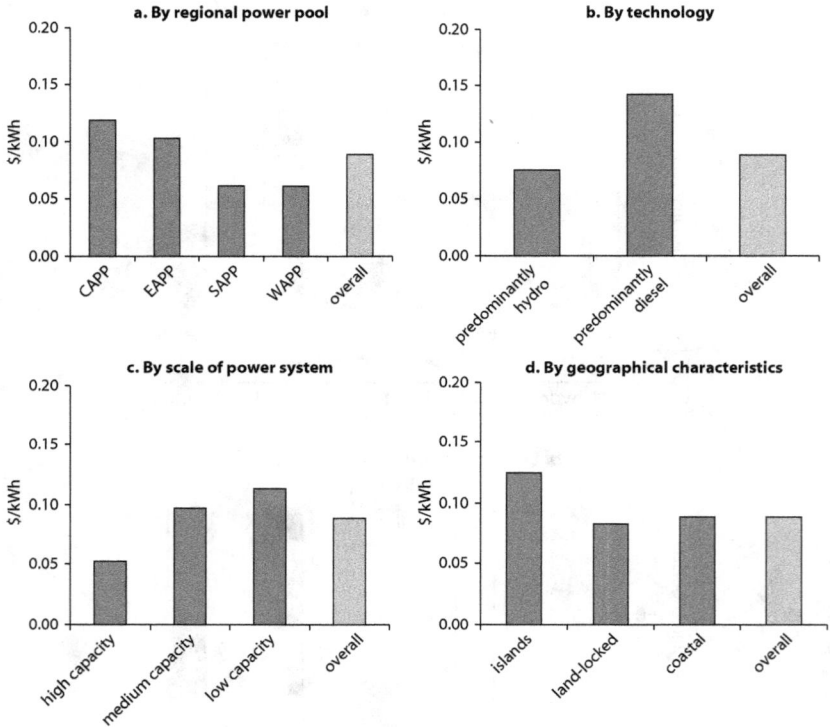

Source: Eberhard and others 2008.
Note: CAPP = Central African Power Pool; EAPP = East African Power Pool; SAPP = Southern African Power Pool; WAPP = West African Power Pool; kWh = kilowatt hour.

installed capacity). Island states face a further cost penalty attributable to the high cost of transporting fossil fuels.

Despite Power Pools, Low Regional Power Trade

Based on the economic geography of the power sector in Sub-Saharan Africa, regional power trade has many potential benefits. In fact, four regional power pools in Sub-Saharan Africa have already been established to promote mutually beneficial cross-border trade in electricity. The theory was that enlarging the market for electric power beyond national borders would stimulate capacity investment in countries with a comparative advantage in generation. The pools would also smooth temporary irregularities in supply and demand in national markets.

Despite high hopes for the power pools, power trade among countries in the region is still very limited. Most trade occurs within the SAPP, largely between South Africa and Mozambique (figure 2.3). Furthermore,

Figure 2.3   Electricity Exports and Imports in Sub-Saharan Africa, 2005

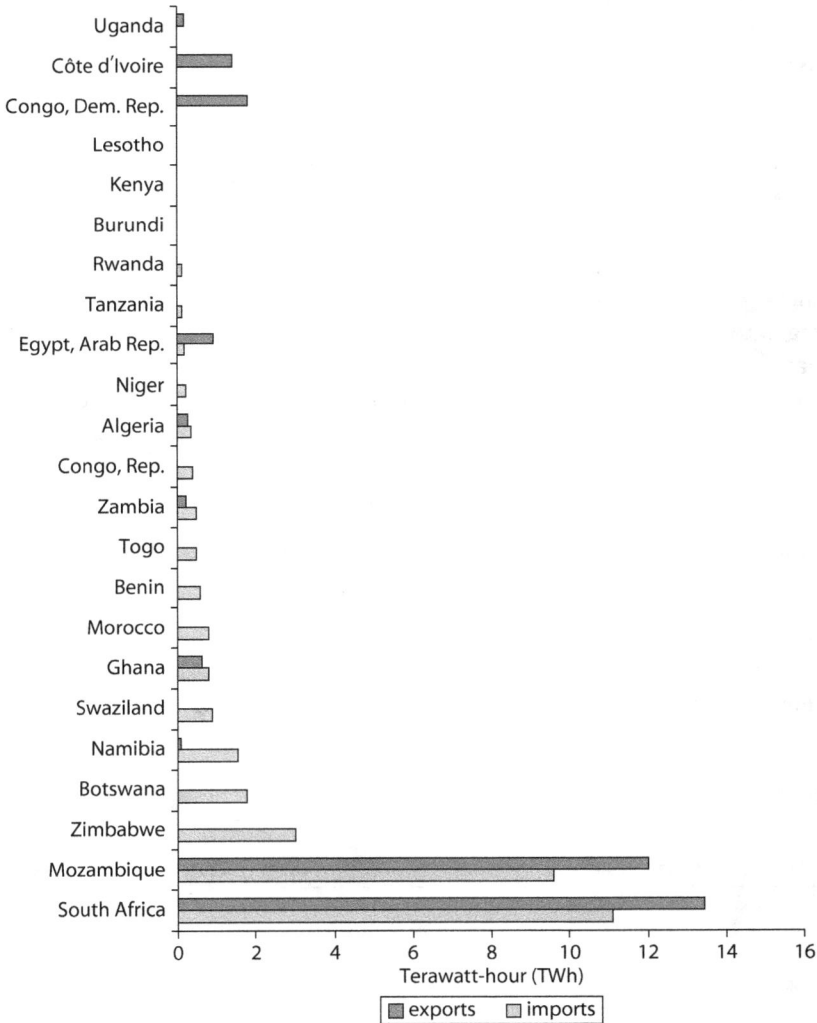

Source: Eberhard and others 2008.
Note: TWh = terawatt-hour.

South Africa reexports much of the electricity it imports from Mozambique back to that country's aluminum smelter.[1] A few countries are highly dependent on imports. In SAPP, Botswana, Namibia, and Swaziland all depend on imports from South Africa. In the West African Power Pool (WAPP, the second-largest pool), Benin, Togo, and Burkina Faso import power from Côte d'Ivoire and Ghana, and Niger imports from Nigeria. The countries of Central Africa engage in minimal power trading, although Burundi, the Republic of Congo, and Rwanda depend on imports from the Democratic Republic of Congo. Power trade in East Africa is negligible.

The region's major exporters generate electricity from hydropower (the Democratic Republic of Congo, Mozambique, and Zambia), natural gas (Côte d'Ivoire and Nigeria), or coal (South Africa). No country that relies on oil or diesel generators exports electricity.

The region's power pools have made progress in developing standard agreements that will allow trade to grow. SAPP has also developed a short-term energy market that enables daily Internet trading. Detailed regulatory guidelines to facilitate cross-border transactions have been prepared by the Regional Electricity Regulators Association (RERA). WAPP also aims to achieve closer regulatory integration in West Africa. Yet despite numerous successes in promoting regional power trade, overall trading volume in the region remains small (table 2.1).

The Potential Benefits of Expanded Regional Power Trading

Rosnes and Vennemo (2008) performed detailed simulations to estimate the potential benefits of regional power trade in Sub-Saharan Africa over a 10-year period from 2005 to 2015. They examine two basic scenarios: *trade stagnation*, in which countries make no further investment in cross-border

Table 2.1   Regional Trade in Electricity, 2005

	Consumption (TWh)	Imports (TWh)	Exports (TWh)	Percentage electricity traded
CAPP	8.80	0.01	1.80	0.1
EAPP	13.41	0.28	0.18	2.1
SAPP	233.97	22.71	25.74	9.7
WAPP	28.63	1.63	2.04	5.7

Source: Eberhard and others 2008.
Note: CAPP = Central African Power Pool; EAPP = East African Power Pool; SAPP = Southern African Power Pool; WAPP = West African Power Pool; TWh = terawatt-hour.

transmission, and *trade expansion*, in which trade occurs whenever the benefits outweigh the costs associated with system expansion. The simulation involved various assumptions regarding input prices, including fuel. To explore the sensitivity of the analysis to changes in assumptions, several subscenarios were considered beyond the base case.

In the trade expansion scenario, annualized power system costs in the trading regions would be 3–10 percent lower. The savings would be the largest in the Central African Power Pool (CAPP) at 10.3 percent, compared with 5–6 percent in SAPP and East African/Nile Basin Power Pool (EAPP/Nile Basin) and only 3.4 percent in WAPP (although savings in some countries in this region are much higher). The annual savings for Sub-Saharan Africa total an estimated $2.7 billion, which is equivalent to 5.3 percent of the annual cost and 7.2 percent of the annual cost when operation of existing equipment is excluded. The savings come largely from substituting hydro for thermal plants, which requires more investment in the short run but substantially reduces operating costs. For example, power trade generates operating cost savings equivalent to 1 percent of regional GDP in EAPP/Nile Basin and almost 0.5 percent of regional GDP in CAPP.

Power trade also reduces the investment requirements of importing countries, which generates further savings. Developing countries, which generally struggle to raise sufficient investment capital to meet their infrastructure needs, clearly benefit from regional power trade.

Under the trade expansion scenario, countries must make additional capital investments to facilitate cross-border transmission. The resulting operating cost savings can therefore be viewed as a substantial return on investment. In SAPP, for example, the additional investment is recouped in less than a year and yields a return of 167 percent. In the other three regions, the additional investment is recouped over three to four years, for a lower—but still generous—return of 20–33 percent. The overall return on trade expansion in Sub-Saharan Africa is 27 percent, which is considerable compared with investments of similar magnitude.

Because trade reduces the use of thermal power plants, the gains from trade increase as fuel prices rise and more hydropower projects become profitable. For example, when the price of oil rises to $75 per barrel (instead of $46 per barrel in the base case), the gains from trade in EAPP/Nile Basin increase from about $1 billion to almost $3 billion.

The 10 largest power importing countries in the trade expansion scenario would reduce their long-run marginal cost (LRMC) of power by $0.02–0.07 per kWh (figure 2.4). Smaller countries that rely on thermal

Figure 2.4   Savings Generated by Regional Power Trade among Major Importers under Trade Expansion Scenario

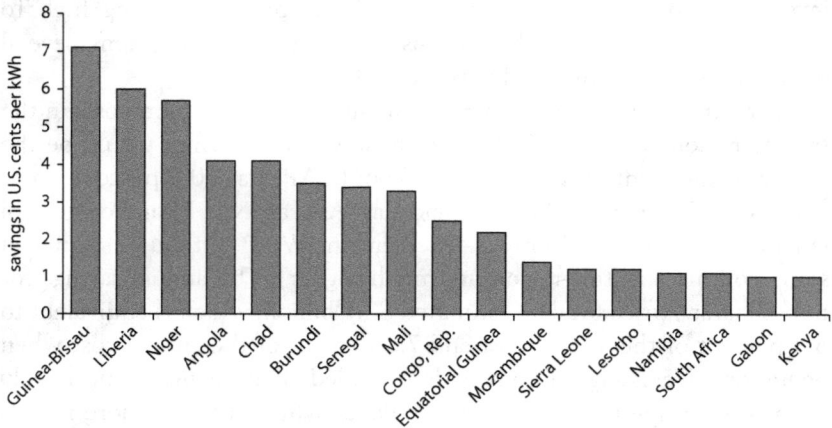

Source: Derived from Rosnes and Vennemo 2008.
Note: kWh = kilowatt hour.

power, such as Burundi, Chad, Guinea-Bissau, Liberia, Niger, and Senegal, stand to gain the most. Nevertheless, reaping the full benefits of power trade will require a political willingness to depend heavily on power imports. As many as 16 African countries would benefit economically by importing more than 50 percent of their power needs.

The future of power trade depends on the health of the power sector in a handful of key exporting countries endowed with exceptionally large and low-cost hydropower resources. In descending order of export potential, these countries are Democratic Republic of Congo, Ethiopia, Guinea, Sudan, Cameroon, and Mozambique (table 2.2). The first three account for 74 percent of the potential exports under trade expansion. Based on a profit margin of $0.01 per kWh, the net export revenue for the top three exporters would account for 2–6 percent of their respective GDP, but the size of the investments to realize these export volumes is daunting. To develop sufficient generation capacity for export, each would need to invest more than $0.7 billion per year, equivalent to more than 8 percent of GDP. Such investments are unlikely to be feasible without extensive cross-border financing arrangements that allow importing beneficiaries to make up-front capital contributions.

Some 22,000 MW of interconnectors would need to be developed to allow power to flow freely across national borders, which would cost

Table 2.2   Top Six Power Exporting Countries in Trade Expansion Scenario

Country	Potential net exports (TWh per year)	Net revenue		Required investment	
		$million per year	% GDP	$million per year	% GDP
Congo, Dem. Rep.	51.9	519	6.1	749	8.8
Ethiopia	26.3	263	2.0	1,003	7.5
Guinea	17.4	174	5.2	786	23.7
Sudan	13.1	131	0.3	1,032	2.7
Cameroon	6.8	68	0.4	267	1.5
Mozambique	5.9	59	0.8	216	2.8

Source: Derived from Rosnes and Vennemo 2008.
Note: GDP = gross domestic product; TWh = terawatt-hour.

more than $500 million a year over the next decade. The return on investment in interconnectors is as high as 120 percent in SAPP and 20–30 percent for the other power pools. For countries with the most to gain from power imports, investments in cross-border transmission have exceptionally high rates of return and typically pay for themselves in less than a year.

What Regional Patterns of Trade Would Emerge?

If regional power trade were allowed to expand, rising demand would provide incentives for several countries to develop their significant hydropower potential. In the trade expansion scenario, for example, the hydropower share of the generation capacity portfolio in SAPP rises from 25 to 34 percent. The Democratic Republic of Congo becomes the region's major exporter of hydropower and exports more than three times its domestic consumption. Mozambique continues to be a significant exporter. Hydropower from the Democratic Republic of Congo flows southward along three parallel routes through Angola, Zambia, and Mozambique (table 2.3 and figure 2.5). Countries such as Angola, Botswana, Lesotho, Malawi, and Namibia subsequently rely on imports to meet more than 50 percent of their power demand. In addition, South Africa continues to import large volumes of power, although imports still account for only 10 percent of domestic consumption.

The EAPP/Nile Basin region experiences a similar shift in generation capacity. The share of hydropower rises from 28 to 48 percent of the generation capacity portfolio, which partially displaces gas-fired power capacity in the Arab Republic of Egypt. Ethiopia and Sudan, the region's

Table 2.3   Power Exports by Region in Trade Expansion Scenario

SAPP	TWh	% Domestic demand	EAPP/Nile basin	TWh	% Domestic demand
Congo, Dem. Rep.	51.9	−369	Ethiopia	26.2	−227
			Sudan	13.1	−13
Mozambique	5.9	−33	Uganda	2.8	−61
Lesotho	−0.7	68	Tanzania	2.4	−22
Malawi	−1.5	56	Rwanda	1.0	−191
Zambia	−1.8	1	Djibouti	0.0	0
Zimbabwe	−3.5	17	Burundi	−0.7	78
Namibia	−3.8	2	Kenya	−2.8	22
Botswana	−4.3	93	Egypt, Arab Rep.	−42.2	32
Angola	−6.0	65			
South Africa	−36.4	10			

WAPP	TWh	% Domestic demand	CAPP	TWh	% Domestic demand
Guinea	17.4	−564	Cameroon	6.7	−84
Nigeria	2.1	−3	Central African Republic	0.0	0
Côte d'Ivoire	0.9	−12			
Gambia, The	0.1	−19	Equatorial Guinea	−0.1	100
Guinea-Bissau	−0.2	7	Gabon	—	42
Mauritania	−0.6	55	Chad	−1.3	102
Benin	−0.9	45	Congo, Rep.	−4.4	4
Sierra Leone	−0.9	60			
Togo	−0.9	48			
Burkina Faso	−1.0	5			
Senegal	−1.4	30			
Niger	−1.5	86			
Liberia	−1.7	89			
Mali	−1.9	79			
Ghana	−9.6	52			

Source: Derived from Rosnes and Vennemo 2008.
Note: CAPP = Central African Power Pool; EAPP/Nile Basin = East African/Nile Basin Power Pool; SAPP = Southern African Power Pool; WAPP = West African Power Pool; TWh = terawatt-hour. — = Not available.

major power exporters, send their power northward into Egypt (see figure 2.5). Exports exceed domestic consumption in both countries. Egypt and Kenya import significant volumes of power (between one-fifth and one-third), but Burundi is the only country to become overwhelmingly dependent on imports (about 80 percent).

Under trade expansion, the share of hydropower in WAPP does not rise significantly. Nevertheless, cost-effective, larger-scale hydropower in

Guinea replaces more dispersed hydropower projects in other countries throughout the region. Gas-fired power plants in Ghana, Benin, Togo, and Mauritania are also avoided—and are replaced by hydropower in Guinea, which emerges as the region's major exporter and exports more than 5 times its domestic consumption.

In CAPP, the share of hydropower increases from 83 percent to 97 percent. Cameroon emerges as the major power supplier in CAPP and exports about half of its production. Hydropower capacity in Cameroon replaces the heavy fuel oil (HFO) –fired thermal capacity in the other countries, in addition to some hydropower in the Republic of Congo. The other countries in the region, except the Central African Republic, import a considerable share of their consumption: Chad and Equatorial Guinea import all of their domestic consumption from Cameroon, and the Republic of Congo imports about one-third of its consumption and Gabon almost half.

Although the benefits of regional power trade are clear, numerous challenges emerge. These are discussed in the remaining sections in this chapter.

Water Resources Management and Hydropower Development

Water resource management for hydropower is challenging for at least two reasons. First, it often requires multinational efforts and joint decision making by several countries. Many rivers with hydropower potential are international. Africa has 60 river basins that are shared by two or more countries, with the largest—the Nile basin—divided among 10 countries. Other important river basins also belong to several states. For example, nine countries share the Niger, eight share the Zambezi, and the Senegal runs through four neighboring states. The development of hydropower capacity therefore depends on the ability of the riparian countries to come to agreements based on joint long-term interests, starting with the location of dams.

Second, hydropower must compete for water resources with other sources of demand: household consumption, irrigation, hydrological regulation, and flood and drought management. Therefore, development of hydropower resources will require an established legal and regulatory framework to facilitate international cooperation and multisectoral management.

Who Gains Most from Power Trade?

Trade is responsible for the substantial differences in the LRMC of power among power pools (table 2.4). For example, in the trade expansion

Figure 2.5   Cross-Border Power Trading in Africa in Trade Expansion Scenario (TWh in 2015)

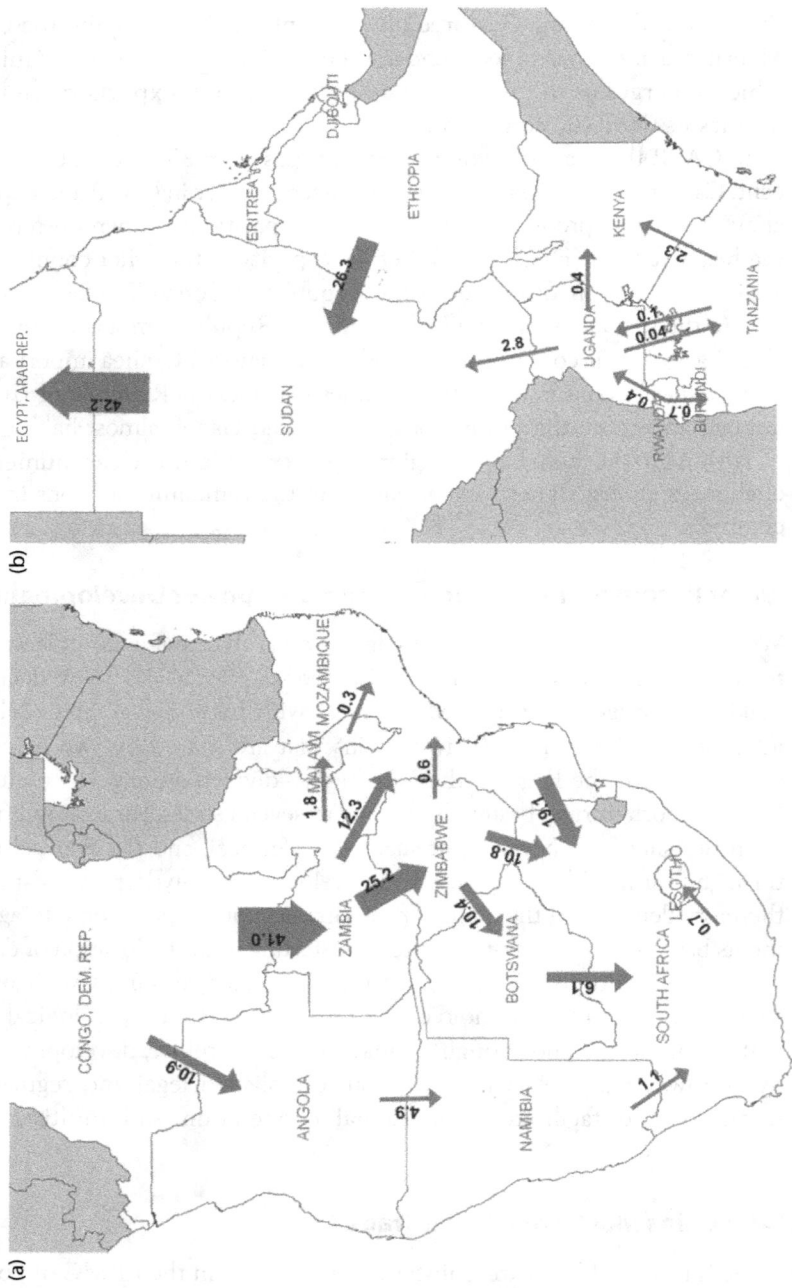

(d)

(c)

Source: Rosnes and Vennemo 2008.
Note: TWh = terawatt-hour.

Table 2.4 Long-Term Marginal Costs of Power under Trade Expansion and Trade Stagnation
$/Kwh

a. SAPP

	Trade expansion	Trade stagnation	Difference
SAPP average	**0.06**	**0.07**	**0.01**
Angola	0.06	0.11	0.05
Botswana	0.06	0.06	0
Congo, Dem. Rep.	0.04	0.04	0
Lesotho	0.06	0.07	0.01
Malawi	0.05	0.05	0
Mozambique	0.04	0.06	0.02
Namibia	0.11	0.12	0.01
South Africa	0.06	0.07	0.01
Zambia	0.08	0.08	0
Zimbabwe	0.08	0.09	0.01

b. EAPP/Nile Basin

	Trade expansion	Trade stagnation	Difference
Basin average	**0.12**	**0.12**	**0**
Burundi	0.11	0.15	0.04
Djibouti	0.07	0.07	0
Egypt, Arab Rep.	0.09	0.09	0
Ethiopia	0.19*	0.16	-0.03
Kenya	0.12	0.13	0.01
Rwanda	0.12	0.12	0
Sudan	0.13	0.13	0
Tanzania	0.10*	0.08	-0.02
Uganda	0.12*	0.11	-0.01

c. WAPP

	Trade expansion	Trade stagnation	Difference
WAPP average	**0.18**	**0.19**	**0.01**
Benin	0.19	0.19	0
Burkina Faso	0.25	0.26	0.01
Côte d'Ivoire	0.15	0.15	0
Gambia, The	0.08	0.07	–0.01
Ghana	0.10	0.10	0
Guinea	0.07	0.06	–0.01
Guinea-Bissau	0.09	0.16	0.07
Liberia	0.08	0.14	0.06
Mali	0.25	0.28	0.03
Mauritania	0.14	0.15	0.01
Niger	0.25	0.30	0.05
Nigeria	0.13	0.13	0
Senegal	0.43	0.47	0.04
Sierra Leone	0.09	0.10	0.01
Togo	0.10	0.11	0.01

d. CAPP

	Trade expansion	Trade stagnation	Difference
CAPP average	**0.07**	**0.09**	**0.02**
Cameroon	0.07	0.06	–0.01
Central African Republic	0.11	0.11	0
Chad	0.07	0.11	0.04

Source: Derived from Rosnes and Vennemo 2008.

Note: CAPP = Central African Power Pool; EAPP/Nile Basin = East African/Nile Basin Power Pool; SAPP = Southern African Power Pool; WAPP = West African Power Pool; kWh = kilowatt-hour.

scenario, the SAPP and CAPP regions have an estimated average LRMC of $0.07 per kWh, which is considerably lower than $0.12 per kWh and $0.18 per kWh for EAPP/Nile Basin and WAPP, respectively. The LRMC varies widely among countries within each power pool, although trade tends to narrow the range.

Trade benefits two types of countries in particular. First, trade allows countries with very high domestic power costs to import significantly cheaper electricity. Perhaps the most striking examples are in WAPP, where Guinea-Bissau, Liberia, and Niger each can save up to $0.06–0.07 per kWh by importing electricity in the trade expansion scenario. Countries in other regions also benefit from substantial savings by importing—up to $0.04–0.05 in Angola in SAPP, Burundi in EAPP/Nile Basin, and Chad in CAPP. Overall savings can be large even for countries with lower unit cost differentials, such as South Africa. Other countries (such as Burundi, Ghana, Malawi, Sierra Leone, and Togo) in the trade expansion scenarios move from being self-reliant to importing heavily, generating savings for each kilowatt-hour that is imported.

Second, expanded trade benefits countries with very low domestic power costs by providing them with the opportunity to generate substantial export revenue. Those countries include Democratic Republic of Congo in SAPP, Ethiopia in EAPP/Nile Basin, Guinea in WAPP, and Cameroon in CAPP. Power export revenue under trade expansion is an estimated 6 percent of GDP in Ethiopia and 9 percent of GDP in the Democratic Republic of Congo. In reality, the parties will need to negotiate terms of trade that will determine the value of exports.

How Will Less Hydropower Development Influence Trade Flows?

In the trade expansion scenario, cheap hydropower from Guinea supplies much of the power in the WAPP region (although not in Nigeria). Realistically, however, it may not be feasible to develop such a huge amount of hydropower in one country and over such a short period. Therefore, in an alternative scenario, only three projects (totaling 375 MW) can be completed in Guinea within the next 10 years (compared with 4,300 MW in the trade expansion scenario).

In this scenario, new trade patterns emerge in the WAPP region. Côte d'Ivoire emerges as the region's major power exporter, and Ghana increases domestic production considerably to reduce net imports. Mauritania and Sierra Leone also become net exporters. Total annualized costs increase by about 3 percent—or just over $300 million—compared with the trade expansion scenario. At the same time, less hydropower is developed to replace thermal capacity, which leads to a huge tradeoff between capital costs and variable costs: Although capital costs are $500 million lower (mainly due to lower generation investments), variable operating costs are $850 million (30 percent) higher. In addition, the existing thermal plants that are used have lower efficiency and higher variable costs than new hydropower capacity.

What Are the Environmental Impacts of Trading Power?

Trade expansion offers potential environmental benefits. In the trade expansion scenario, the share of hydropower generation capacity in SAPP rises from 25 to 34 percent, reducing annual carbon dioxide emissions by about 40 million tons. Power production rises by 2.4 TWh in the EAPP/Nile Basin region, yet carbon dioxide emissions still fall by 20 million tons. Reduction in thermal capacity is smaller in WAPP and CAPP, and emissions savings are correspondingly lower: 5.2 and 3.6 million tons, respectively.

The International Energy Agency recently estimated that emissions from power and heat production in Africa are 360 million tons. Under the trade expansion, carbon dioxide emissions fall by 70 million tons, or 20 percent of total emissions. These estimates do not, however, include greenhouse gas emissions from hydropower in the form of methane from dams.

Technology Choices and the Clean Development Mechanism

The Clean Development Mechanism (CDM) allows industrialized countries that have made a commitment under the Kyoto protocol to reduce greenhouse gases to invest in projects that reduce emissions in developing countries. The CDM facilitates financing to cover the difference in cost between a polluting technology and a cleaner but more expensive alternative. The cost of certified emission reduction credits (CERs) associated with a given project is calculated by dividing the difference in cost by the resulting reduction in emissions. Rosnes and Vennemo (2008) analyze the potential for CDM in the power sector in SAPP under the trade-expansion scenario.

Based on a CER price of $15 per ton of carbon dioxide, the CDM stimulates investments in the Democratic Republic of Congo, Malawi,

Namibia, and Zambia and adds 8,000 MW (producing 42 TWh) of hydropower capacity.

At the same CER price, the CDM has the potential to reduce carbon dioxide emissions in SAPP by 36 million tons, equivalent to 10 percent of the continent's current emissions from power and heat production. Although significant, that is still less than the carbon reduction brought about by trade, which reduces emissions by 40 million tons in SAPP. Trade and CDM are not mutually exclusive, of course. Compared with trade stagnation, trade expansion combined with the CDM generates emissions reductions of 76 million tons.

How Might Climate Change Affect Power Investment Patterns?

Because unpredictable weather patterns reduce hydropower's reliability, climate change could increase the costs of generating and delivering power in Africa. Rosnes and Vennemo (2008) therefore performed a simulation to estimate the effect of climate change on costs in EAPP/Nile Basin. They assumed that climate change affects both existing and new capacity, reducing hydropower production (in gigawatt-hours per megawatt of installed capacity) by up to 25 percent.

Lower firm power would increase the unit cost of hydropower, cause gradual substitution away from hydropower, and increase the total annualized cost of the power sector. In this scenario, a 25 percent reduction in firm hydropower availability would increase the annual costs of meeting power demand by a relatively low 9 percent. At the same time, however, reliance on thermal power would increase by 40 percent in EAPP/Nile Basin. In other words, climate change is a sort of positive feedback loop: Sustainable power becomes less reliable and therefore more expensive. It leads to increased reliance on thermal power, which exacerbates the climate problem.

Meeting the Challenges of Regional Integration of Infrastructure

Increased regional power trade in Africa has clear benefits. Developing sufficiently integrated regional infrastructure, however, poses substantial political, institutional, economic, and financial challenges for policy makers. The first step to meeting those challenges is to build political consensus among neighboring states that may have diverging national agendas or even recent histories of conflict. Thereafter, effective regional institutions will be needed to coordinate a cross-border infrastructure development program and ensure an equitable distribution of benefits. Power needs in the region are vast, but resources are limited. Policy makers will therefore need to set priorities to guide regional integration. Even with clear priorities, however, funding and implementing extensive project preparation studies and arranging cross-border finance for complex, multibillion-dollar projects present considerable difficulties. The efficacy of regional infrastructure will ultimately depend on countries to coordinate associated regulatory and administrative procedures (box 2.1).

Building Political Consensus

Developing appropriate regional infrastructure is only one aspect of regional integration. Compared with economic or political integration, infrastructure integration has more clearly defined benefits and requires countries to cede less sovereignty. Regional infrastructure cooperation is therefore a good first step toward broader integration.

Some countries have more to gain from regional integration than others. In particular, regional power trade benefits small countries with high power costs. As long as regional integration provides substantial economic advantages, however, it should be possible to design compensation mechanisms that benefit all participating countries. Benefit sharing was pioneered through international river basin treaties and has applications for integration of regional infrastructure.

Any regional initiative requires national and international political consensus. Methods for building consensus vary, but broad principles apply.

Improved advocacy. Africa will require improved high-level advocacy and leadership to promote regional integration for infrastructure development. Regional integration issues remain only a small part of parliamentary debate in most countries. The infrequency of regional meetings of heads of state contributes to a lack of follow-through. Governments and international institutions must therefore provide leadership. The African Union (AU) has the mandate to coordinate the regional integration program defined by the 1991 Abuja Treaty, which created the African Economic Community with regional economic communities as building blocks. The New Partnership for Africa's Development (NEPAD) is the main vehicle for promoting regional integration but so far has not received sufficient support from political leaders to build consensus around financially and economically viable projects. The NEPAD Heads

Box 2.1

The Difficulties in Forging Political Consensus: The Case of Westcor

On October 22, 2004, the Energy Ministers of Angola, Botswana, the Democratic Republic of Congo, Namibia, and South Africa signed an Intergovernmental Memorandum of Understanding pledging cooperation on two projects: the establishment and development of the third phase of the Inga hydroelectric program in the Democratic Republic of Congo and the power export from there to the other four countries via a new Western Power Corridor transmission system. The chief executives of the five national utilities signed a similar memorandum of understanding among themselves. The Westcor company was established in September 2005 to take the project forward. It is registered in Botswana and has equal shareholdings by the five participating countries.

Inga 3 was expected to deliver 3,500 MW. Additional hydroelectric plants in Angola and Namibia were also seen as possibilities. Inga is one of the most favorable hydro sites in the world. It is situated in the rapids coursing around a U-shaped bend in the massive Congo River. By cutting through the peninsula, a run-of-river hydroelectric operation can be developed without the construction of massive storage dams. Inga 1 (354 MW) and Inga 2 (1424 MW) were built many years ago and are being rehabilitated.

A prefeasibility study was completed that suggested potentially attractive power costs. A detailed design was originally scheduled for 2008–09. Despite intensive political lobbying within the African Union, New Partnership for Africa's Development, Southern African Development Community, Southern African Power Pool, and development finance institutions, funds have yet to be committed to conduct a full feasibility study. There are also considerable obstacles to the conclusion of regulatory, contractual, and financing agreements.

In 2009, the government of the Democratic Republic of Congo announced that it was negotiating with BHP Billiton to assist in the development of Inga 3, including a large investment in an aluminum smelter that would be the main offtaker for the project. Westcor has subsequently closed its project office. In the absence of political consensus and meaningful commitment, the future of hydroelectric exports from Inga remains uncertain.

Source: Interviews conducted by the authors with staff in the Africa Energy Department of the World Bank, 2009.

of State Implementation Committee, established to remove political obstacles to projects, has not been effective and now meets less regularly than originally. A strong commitment from regional leaders is therefore essential to move projects forward. For example, when political differences threatened to derail the West Africa Gas Pipeline, only the shuttle diplomacy of Nigeria's President Obasanjo kept the project on track.

Stronger trust. Trust is important for regional integration—especially when some countries stand to benefit more than others. Countries may be able to build that trust by collaborating on small, well-defined projects. For example, a bilateral agreement for a cross-border power transaction may be easier to conclude than a large regional investment that requires multicountry off-take agreements. Frequent interaction among policy makers at all levels of government builds relationships that help overcome inevitable disagreements. Finally, supranational organizations can serve as honest brokers for sharing gains and resolving disputes.

Credible information. Trust is easier to build when information is shared equally. Decision makers require accurate data to gauge the full costs and benefits of regional infrastructure investments, many of which involve allocating substantial funds and sacrificing some degree of sovereignty. Regional economic communities are then responsible for building consensus by ensuring that all stakeholders are aware of the potential benefits of investments. Otherwise, countries are unlikely to be willing to bear the full cost of public goods. A realistic and accurate assessment of the likely benefits and costs of regional integration will therefore help to build trust among countries.

Strengthening Regional Institutions

Africa has many regional institutions, but most are ineffective. The architecture supporting African integration comprises more than 30 institutions, including executive continental bodies, regional economic communities with overlapping membership, sectoral technical bodies, and national planning bodies. As a result, it is unclear who is responsible for strategy planning, project development, and financing. This has slowed the development of cohesive regional strategies, establishment of realistic priorities (such as regional infrastructure and trade integration), and design of technical plans for specific projects.

The AU Commission has struggled to fulfill its mandate because of a lack of human and financial resources. Africa's regional economic communities have limited capabilities and resources and, above all, weak authority to enforce decisions. Institutions would be more effective if governments were willing to cede a measure of sovereignty in return for greater economic benefits. Greater use of qualified majority rules (which has been an issue of debate for some time in many regional economic communities, although without resolution) in some areas of policy making would streamline decision making. Furthermore, member states often fail to pay their assessed contributions in full, which constrains financing. Regional economic communities have multiple functions, and infrastructure provision is not always at the forefront (ICA 2008). As a result, they often fail to attract and retain professional staff with the experience to identify and promote complex regional infrastructure projects.

Regional special purpose entities or sectoral technical bodies—such as power pools—have been more effective than regional economic communities. A power pool has a clear mandate, sufficient autonomy to execute its responsibilities, a dedicated funding mechanism, and career opportunities that attract and retain high-caliber staff. It also receives substantial capacity building. The members of a power pool are national electricity utilities, which similarly have clear functions and roles within their national contexts and are less susceptible to immediate political pressures than are less technical public agencies.

Some power pools have been more proactive in promoting the development of their power sector. For example, WAPP appears to be taking initiative in promoting investment and assisting in the establishment of a regional electricity regulator (box 2.2). By contrast, SAPP, despite a longer history, seems more concerned with protecting the interests of its member national utilities than with facilitating the entry of private investment.

National agencies are also in need of capacity building and streamlined decision making. For complex regional infrastructure projects, several line ministries from each country are often involved, which complicates consensus building and obscures responsibilities. High-level government officials often fail to implement regional commitments.

Setting Priorities for Regional Infrastructure

The financial distress of many utilities in Africa has resulted in a substantial backlog of infrastructure investment. Authorities in Africa must therefore set effective investment priorities, especially considering the limited fiscal space and borrowing ability of many governments. Because infrastructure has a long life, unwise investments can burden governments with an ineffective project that will also require costly maintenance.

Box 2.2

The West African Power Pool (WAPP) and New Investment

Unlike other power pools in Africa, WAPP is responsible for developing new infrastructure. The WAPP Articles of Association require WAPP to ensure "the full and effective implementation of the WAPP Priority Projects."

The WAPP Executive Board is responsible for developing a regional transmission and generation master plan. Within the WAPP Secretariat, the Secretary General negotiates directly with donors to finance feasibility studies for new projects and subsequently secures grant financing for feasible projects. WAPP has already obtained funding for feasibility studies from several donors, including the World Bank and U.S. Agency for International Development.

WAPP often works with multilateral development banks to secure grant or credit financing for development projects. For example, grants and credits from the World Bank and KfW account for all funding of investments for the Coastal Transmission Backbone. In other cases, WAPP has created a special purpose vehicle that allows members to take equity stakes in projects, including a number of regional hydro generation projects.

Source: Castalia Strategic Advisors 2009.

Although our modeling has indicated clear overall benefits for expanded regional trade, many large regional projects are difficult to develop: Financing sums are large, policy and regulatory environments are diverse, and agreements have to be forged between affected stakeholders. Some observers may argue that it is easier to begin by developing smaller national projects that have lower financing requirements and less complex regulatory and decision-making environments. However, these may be more costly in terms of power generated. Therefore, it still makes sense to prioritize regional projects and first develop those that have the highest economic returns and still have a reasonable chance of reaching financial closure.

For many years, regional power pools have been developing regional power plans with lists of possible projects. Yet they have struggled to agree on priorities: All members want their pet projects on the short list, and national utilities have also been protective of their market dominance (box 2.3).

Suitable criteria for priority projects include predicted economic returns and scope for private participation.

Box 2.3

Difficulties in Setting Priorities in SAPP

In Southern Africa, energy ministers from the Southern African Development Community (SADC) asked the Southern African Power Pool (SAPP) to prepare a priority list for power projects in the region. SAPP, in turn, asked utilities to provide information on the power projects located in their area. By late 2005, SAPP had prepared a priority list based on seven weighted criteria: project size, leveled energy cost, transmission integration, economic impact, percentage of offtake committed, regional contribution, and number of participating countries. Projects were divided into four categories: rehabilitation, transmission, short-term generation, and long-term generation. SAPP presented the priority list to SADC energy ministers, but they failed to reach an agreement. Individual ministers generally favored projects located in their country, and inevitably some countries had a less significant presence on the list. SAPP then presented an amalgamated list of all possible power projects in the region at an investor conference in 2007. SAPP failed to demonstrate the necessity and viability of the projects, and as a result none of them received financing. Having twice failed to design an acceptable priority list, SAPP hired consultants to prepare a least-cost pool plan and prioritize projects. The recommendations were again controversial, and SAPP failed to achieve consensus on the priority list.

With the region still in need of infrastructure investment, a group that included SADC, SAPP, Development Bank of Southern Africa, and RERA (the Regional Electricity Regulatory Authority of the Economic Community of West African States) asked consultants to prepare a list of short-term regional power projects that required financing. The focus of this list was on getting bankable projects, given that most utilities within SAPP cannot support the required investments on their balance sheet. The consultants sought projects that met four criteria: financial close within 24 months, least-cost rationale, regional impact, and environmental considerations. Developers and project sponsors presented the final list at an investor conference in mid-2009, but none of the projects has yet reached financial close.

Source: Interviews conducted by the authors with staff in the Africa Energy Department of the World Bank, 2009.

Economic returns. Projects with the highest returns may not always be new infrastructure. Strategic investments that improve the performance of existing infrastructure systems, such as installing power interconnectors between countries with large cost differentials, are often the most cost effective.

Scope for private participation. The prospect of a larger regional market can attract more interest for private financing and public-private partnerships, which provides a possible solution to the region's substantial financing gaps. Encouraging private sector involvement requires government cooperation to facilitate investment. In fact, public control in many countries continues to stifle private investment. For many years, the membership of power pools, such as SAPP, was restricted to state-owned national utilities. The rules have changed, but independent power projects still face many obstacles to gaining full membership in power pools.

Priority-setting exercises are under way or planned. For example, a joint AU–African Development Bank study, the Program for Infrastructure Development in Africa, aims to develop a vision of regional infrastructure integration on the continent. The study will need to take account of other ongoing processes such as the Africa–European Union Energy Partnership, which is working to gain consensus on an electricity master plan for Africa. In addition, many regional economic communities and other technical regional institutions have 10-year investment plans that provide many opportunities for external financiers.

Priority setting depends on transparency in decision making and agreement on selection criteria. Decisions must be based on sufficiently detailed data and reasonable assumptions, and results should be publicly available. Small investments in better information at the country and regional levels will have significant benefits for decision making, especially given the size of public and private funds at stake.

Facilitating Project Preparation and Cross-Border Finance

Project design is a complex process. The appraisal phase establishes social, economic, financial, technical, administrative, and environmental feasibility (Leigland and Roberts 2007). For regional projects, coordination among national agencies with different procedures, capacity, and administrative constraints adds to the complexity. As a result, the project preparation costs for regional projects tend to be higher, and the process can take longer than for national projects.

Preparation costs for regional projects are typically around 5 percent of total financing—approximately double the cost of preparing national projects. These costs are incurred when the success of the project and the likelihood of a sufficient return from the investment are still uncertain. Regional institutions and donors have tried to address these challenges and have established more than 20 project preparation facilities, many of which explicitly support regional activities. Unfortunately, available project preparation resources do not match the regional needs. African countries need to commit more funds and people with the right technical, legal, and financial skills for infrastructure planning and project implementation. Timely execution of project preparation activities and a steady supply of new projects also encourage participation of the private sector. For operators relying on private financing, a firm planning horizon is therefore even more critical than for the public sector.

Multilateral institutions have been developing specific mechanisms for funding regional projects. The World Bank has five criteria for regional projects to qualify for concessional funding from the International Development Association (IDA): At least three countries must participate, although they can enter at different stages; countries and the relevant regional entity must demonstrate strong commitment; economic and social benefits must spill over country boundaries; projects must include provisions for policy coordination among countries; and projects must be priorities within a well-developed and broadly supported regional strategy. A recent evaluation of World Bank regional integration projects concluded that regional programs have been effective (World Bank 2007).

The African Development Bank adopted similar principles in 2008, although requiring only two countries to participate. To encourage greater country ownership, both institutions use a one-third, two-thirds principle, whereby participants are expected to use one IDA or African Development Fund credit from their country allocation, supplemented by two credits from regionally dedicated resources. Currently 17.5 percent of the African Development Fund and 15 percent of IDA resources in Africa are dedicated to regional programs. For projects to be eligible for financing from the European Union–Africa Infrastructure Trust Fund, they must be sustainable and have African ownership. They must also be cross-border projects or national projects with a regional impact on two or more countries. Regional projects funded by the Development Bank of Southern Africa must either involve a minimum of two countries or be located in a single country with benefits to the region.

Small, poor countries with the potential to develop large hydro projects supplying multiple countries face considerable obstacles in financing these projects. For example, the countries must sign secure power purchase agreements with large power loads to provide predictable revenue streams. Large, financially viable utilities; industrial customers in neighboring countries; or new adjacent energy-intensive investments, such as aluminum smelters, are potential sources for anchor loads, but they are not always available. The alternative is to combine multiple cross-border power off-take agreements, which will be challenging.

Further challenges remain. Although recipients of funds from the African Development Fund and the IDA can leverage their country allocations by participating in regional projects, those receiving a small allocation may be reluctant to use a large percentage on one regional project with unclear benefits. How such concessional resources are allocated and whether enough of the overall allocation is dedicated to regional projects remain issues of debate. In addition, development finance institutions offer limited financing instruments for middle-income countries. This is problematic for projects involving Botswana and South Africa as well as North Africa, which could benefit from connectivity with countries south of the Sahara.

IDA guidelines do not permit grants to regional organizations or supranational projects. This limits the World Bank's ability to provide capacity building for weak regional agencies. Some projects with significant regional spillovers—such as the Ethiopia-Sudan interconnector and a thermal power generation project in Uganda—may not involve three or more countries and therefore do not qualify for concessionary regional financing.

Developing Regional Regulatory Frameworks

Physical infrastructure will not produce economic growth on its own. To ensure its efficient use, the legal, regulatory, and administrative environment must be improved. Worldwide experience in developing power pools has led to consensus on three key building blocks for success: a common legal and regulatory framework, a durable framework for systems planning and operation, and an equitable commercial framework for energy exchanges.

Political, regulatory, and physical barriers limit power trade—and therefore market size—throughout Africa. Regional power infrastructure requires coordinated power pricing, third-party access regulations, and effective cross-border trading contracts.

The four power pools in Sub-Saharan Africa are at different stages of development. As countries move from bilateral to multilateral power exchanges, however, a commercially acceptable framework will be essential. The WAPP was granted special status by the Economic Community of West African States (ECOWAS) in 2006 to reinforce its autonomy, and the 2007 ratification of an overarching Energy Protocol will help promote security for investors and open access to national transmission grids across the region. In 2008 the ECOWAS Regional Electricity Regulatory

Authority was established to regulate cross-border electricity exchanges between member states.

In Southern Africa, RERA has developed guidelines for cross-border power projects. These were formally noted by a Southern African Development Community (SADC) meeting of energy ministers in April 2010. RERA is now disseminating the guidelines among its member regulatory agencies.

Conclusion

Cross-border trade in power has significant potential to lower costs and stimulate investment. In the short run, greater investments in cross-border transmission links will be needed to accommodate the higher volume of trade, but those investments would be quickly repaid as countries gain access to cheaper power, particularly in Southern Africa. Although the overall savings in the annualized cost of the power sector under trade are relatively small (less than 10 percent), the gains for individual countries may be substantial. Development finance institutions should consider accelerating investments in cross-border transmission links and large hydroelectric projects, which the private sector has found too risky because of their high capital costs, long payback periods, and risks related to the enforceability of power-purchase agreements.

Note

1. Investment in the large Cahora Bassa hydroelectric plant in Mozambique was justified on the basis of exports of electricity to South Africa. Subsequently, South Africa had excess generation capacity that was made available for a new aluminum smelter built in the port city of Maputo.

Bibliography

AICD (Africa Infrastructure Country Diagnostic). 2008. *AICD Power Sector Database*. Washington, DC: World Bank.

Castalia Strategic Advisors. 2009. "International Experience with Cross Border Power Trading. A Report to the ECOWAS Regional Electricity Regulatory Authority." http://www.esmap.org/esmap/sites/esmap.org/files/P111483_AFR_International%20Experience%20with%20Cross-Border%20Power%20Trading_Hughes.pdf.

Eberhard, Anton, Vivien Foster, Cecilia Briceño-Garmendia, Fatimata Ouedraogo, Daniel Camos, and Maria Shkaratan. 2008. "Underpowered: The State of the

Power Sector in Sub-Saharan Africa." Background Paper 6, Africa Infrastructure Country Diagnostic, World Bank, Washington, DC.

ICA (Infrastructure Consortium for Africa). 2008. "Mapping of Donor and Government Capacity-Building Support to African RECs and Other Regional Bodies." Report of Economic Consulting Associates to the Infrastructure Consortium for Africa, Tunis.

Leigland, James, and Andrew Roberts. 2007. "The African Project Preparation Gap: Africans Address a Critical Limiting Factor in Infrastructure Investment." PPIF Note, World Bank, Washington, DC.

Rosnes, Orvika, and Haakon Vennemo. 2008. "Powering Up: Costing Power Infrastructure Spending Needs in Sub-Saharan Africa." Background Paper 5, Africa Infrastructure Country Diagnostic, World Bank, Washington, DC.

World Bank. 2007. *The Development Potential of Regional Programs: An Evaluation of World Bank Support of Multicountry Operations.* Washington, DC: World Bank, Independent Evaluation Group.

CHAPTER 3

Investment Requirements

Meeting Africa's infrastructure needs will require substantial investment. Projections of future physical infrastructure requirements provide the basis for estimates of spending requirements in this chapter. In all cases, the spending estimates account for both growth-related and social demands for infrastructure and maintenance and rehabilitation costs.

We assume that over a 10-year period the continent should be expected to redress its infrastructure backlog, keep pace with the demands of economic growth, and attain a number of key social targets for broader infrastructure access. In this chapter, potential generation projects in the Central, East/Nile Basin, Southern, and West African power pools (CAPP, EAPP/Nile Basin, SAPP, and WAPP, respectively) are identified and ranked according to cost effectiveness.

Installed capacity will need to grow by more than 10 percent annually— or more than 7,000 megawatts (MW) a year—just to meet Africa's suppressed demand, keep pace with projected economic growth, and provide additional capacity to support efforts to expand electrification. In the decade before 2005, expansion averaged barely 1 percent annually, or less than 1,000 MW per year. Most new capacity would be used to meet nonresidential demands from the commercial and industrial sectors.

Based on these assumptions, the overall costs for the power sector between 2005 and 2015 in Sub-Saharan Africa are a staggering $41 billion a year—$27 billion for investment and $14 billion for operations and maintenance. Development of new generating capacity constitutes about half of investment costs, and rehabilitation of existing generation and transmission assets about 15 percent. SAPP alone accounts for about 40 percent of total costs.

Modeling Investment Needs

Nowhere in the world is the gap between available energy resources and access to electricity greater than in Sub-Saharan Africa. The region is rich in oil, gas, and hydropower potential, yet more than two-thirds of its population lacks access to electricity. Coverage is especially low in rural areas. National authorities and international organizations have drawn up plans to increase access, but policy makers must make key decisions to underpin these plans, such as how rapidly the continent can electrify, which mode of power generation is appropriate in each setting, and whether individual countries should move ahead independently or aim for coordinated development. They must also realistically assess the effect of major global trends, such as rising oil prices and looming climate change, their impact on decision making, and the sensitivity of power investment decisions to broader macroeconomic conditions.

To inform decision making, Rosnes and Vennemo (2008), as part of the Africa Infrastructure Country Diagnostic study, developed a model to analyze the costs of expanding the power sector over the course of 10 years under different assumptions. The model simulates optimal (least cost) strategies for generating, transmitting, and distributing electricity in response to demand increases in each of 43 countries participating in the four power pools of Sub-Saharan Africa: the Southern African Power Pool, the East African/Nile Basin Power Pool,[1] the West African Power Pool, and the Central African Power Pool.[2] Cape Verde, Madagascar, and Mauritius are also included in our study as island states. Each power pool has dominant players. For example, South Africa accounts for 80 percent of overall power demand in SAPP, the Arab Republic of Egypt for 70 percent in EAPP/Nile Basin, Nigeria for two-thirds in WAPP, and the Republic of Congo and Cameroon for a combined 90 percent of power demand in CAPP.

The cost estimates are based on projections of power demand over the 10 years between 2005 and 2015. Demand has three components: *market demand* associated with different levels of economic growth, structural change, and population growth; *suppressed demand* created by frequent blackouts and the ubiquitous power rationing; and *social demand*, which is based on political targets for increased access to electricity.

In most low-income countries, notional demand exceeds supply.[3] The difference between the two is suppressed demand, which arises for two primary reasons. First, people who are on a waiting list to get connected are not captured in baseline demand estimates. Second, frequent blackouts and brownouts reduce consumption but not notional demand. Ultimately, suppressed demand will immediately absorb a certain amount of new production even before taking account of income growth or structural economic changes.

In their model, Rosnes and Vennemo (2008) account for suppressed demand differently depending on its source. Waiting lists are a direct result of slow connection and expansion, and so they assume that social demand will include suppressed demand from this source in each scenario. Suppressed demand from blackouts, on the other hand, is estimated based on data for blackout duration and frequency from the World Bank's enterprise surveys (table 3.1). They then adjust electricity demand in the base year (2005) accordingly.

Social demand for electricity includes the expected demand of all new connections in the household sector in 2015 (table 3.2). Rosnes and Vennemo (2008) examine three scenarios for electricity access. In the *constant access scenario*, access rates remain at their 2005 level. Because of population growth, even the constant access scenario implies a number of new connections and therefore greater demand in kilowatt-hours (kWh). In the *regional target access scenario*, access rates increase by roughly one percentage point per year in each region—an ambitious but still realistic target. Finally, in the *national targets scenario*, access rates reflect targets set by national governments for urban and rural electricity access.

Based on historic trends, demand is projected to grow at 5 percent per year in Sub-Saharan Africa and reach 680 terawatt-hours (TWh) by 2015. In all scenarios, market demand accounts for the great bulk of demand growth over the period.

Estimating Supply Needs

To estimate supply, the model simulates the least expensive way of meeting projected demand. Calculations are based on cost assumptions for various investments, including refurbishment of existing capacity

Table 3.1   Blackout Data for Selected Countries

	Outages (days/year)	Average duration (hours)	Outages (hours per year)	Down time (% of year)	Suppressed demand in 2005 (GWh)
Southern African Power Pool					
Angola	92	19.31	1,780.8	20.3	435
Congo, Dem. Rep.	182	3.63	659.2	7.5	351
South Africa	6	4.15	24.5	0.3	602
Zambia	40	5.48	219.9	2.5	157
East African/Nile Basin Power Pool					
Kenya	86	8.20	702.6	8.0	366
Tanzania	67	6.46	435.9	5.0	208
Uganda	71	6.55	463.8	5.3	84
Western African Power Pool					
Côte d'Ivoire	46	5.94	1,101	13	365
Ghana	61	12.59	1,465	17	979
Nigeria	46	5.94	1,101	64	10,803
Senegal	44	5.67	1,052	17	250
Sierra Leone	46	5.94	1,101	82	189
Central African Power Pool					
Cameroon	26	4.03	613	7.0	241
Congo, Rep.	39	4.33	924	10.6	616
Gabon	40	5.20	950	10.8	134

Source: Rosnes and Vennemo 2008.
Note: GWh = gigawatt-hour.

Table 3.2   Projected Market, Social, and Total Net Electricity Demand in Four African Regions
TWh

Region	Total net demand 2005	Market demand 2015	Social demand 2015	Total net demand 2015	Annual average growth rate (%)
SAPP	258.8	383.0	14.0	397.0	4.4
EAPP/Nile Basin	100.6	144.8	24.2	169.0	5.3
WAPP	31.3	69.6	24.9	94.5	11.8
CAPP	10.7	17.0	3.0	20.0	6.7
Total	401.4	614.4	66.1	680.5	5.5

Source: Rosnes and Vennemo 2008.
Note: Social demand is based on national connection targets. CAPP = Central African Power Pool; EAPP = East African/Nile Basin Power Pool; SAPP = Southern African Power Pool; WAPP = West African Power Pool; TWh = terawatt-hour.

for electricity generation and construction of new capacity for cross-border electricity transmission. The model includes four modes of thermal generation—natural gas, coal, heavy fuel oil, and diesel—and four renewable generation technologies—large hydropower, mini-hydro, solar photovoltaic, and geothermal. Operation of existing nuclear capacity is also considered, although new investment is not.

Initial supply is based on the existing generation capacity in the base year of 2005. Expansion is possible through investments in both new capacity and refurbishment of existing capacity to extend its life. The investment costs for each technology include both capital and variable operating costs (including fuel and maintenance). Expanding access will also require investment to extend and refurbish the transmission and distribution (T&D) grid and enhance off-grid options; these will also require maintenance.

The model can be run under a number of scenarios with varying assumptions to highlight the policy implications of each. As mentioned previously, for example, the feasibility of meeting three different electrification targets in each region is examined (table 3.3). A *lower growth scenario* assumes lower gross domestic product (GDP) growth. To assess the effect of trade on investment and operating costs, two trade scenarios were simulated. In the *trade expansion* scenario, trade will expand wherever it is worth the cost—that is, wherever the benefits of trade outweigh

Table 3.3   Projected Generation Capacity in Sub-Saharan Africa in 2015 in Various Scenarios
MW

Generation capacity (MW)	Trade expansion scenario			Trade stagnation scenario	Low-growth scenario
	Constant access rate	Regional target access rate	National targets for access rates	National targets for access rates	National targets for access rates, trade expansion
Installed capacity[a]	43,906	43,906	43,906	43,906	43,906
Refurbished capacity	35,917	36,561	37,382	37,535	35,945
New capacity	74,366	77,953	81,722	70,425	65,723

Source: Adapted from Rosnes and Vennemo 2008.
Note: MW = megawatt.
a. "Installed capacity" refers to installed capacity as of 2005 that is not refurbished before 2015. Existing capacity that is refurbished before 2015 is included in the "refurbished capacity."

the costs of the additional infrastructure needed to support expanded trade. In another scenario—*trade stagnation*—no further investment in cross-border grids is made. The model has guidelines for endogenously determining trade flows, which can increase (in the trade expansion scenarios) or even switch direction compared with the 2005 trade pattern.

To meet national electrification targets in 2015 under the trade expansion scenario, the region will need about 82,000 MW of new generation capacity—almost equal to total capacity in 2005.

Because many power installations in Africa are old, much of the capacity operating in 2005 will need to be refurbished before 2015. The 2005 capacity in SAPP was 48,000 MW. Approximately 28,000 MW of generation capacity will have to be refurbished by 2015. In addition, the region requires more than 33,000 MW of new generation capacity, an increase of about 70 percent over 2005 capacity. EAPP/Nile Basin has minimal refurbishment needs but requires 17,000 MW of new capacity— approximately equal to the region's installed capacity in 2005. New capacity requirements in WAPP and CAPP are also significant: 18,000 MW in WAPP, or 180 percent of 2005 capacity, and 4,400 MW in CAPP, or 250 percent of 2005 capacity. More than half of 2005 capacity must be refurbished in both WAPP and CAPP—7,000 and 900 MW, respectively.

Investment requirements are challenging in every region, although they are particularly large in WAPP and CAPP. Fortunately, however, the model's projections indicate that economic growth will drive most of the growth in demand. Therefore, each region's financial strength will grow to meet new investment needs as they arise.

Table 3.4 provides a summary of new connections that will need to be made to meet national electrification targets by 2015 in the different regions.

Overall Cost Requirements

The overall costs for the power sector in Africa (including Egypt) between 2005 and 2015 (based on the trade expansion scenario and national targets for access rates) are an estimated $47.6 billion a year— $27.9 billion for investment and $19.7 billion for operations and maintenance (table 3.5).

About half of the investment cost is for development of new generation capacity and another 15 percent for rehabilitation of existing generation and transmission assets. SAPP alone accounts for about 40 percent of costs.

**Table 3.4   New Household Connections to Meet National
Electrification Targets, 2005–15**

Pool	New household connections (millions)
CAPP	2.5
EAPP/Nile Basin	20.0
SAPP	12.2
WAPP	21.5
Island states[a]	1.2
Total	57.4

Source: Adapted from Rosnes and Vennemo 2008.
Note: CAPP = Central African Power Pool; EAPP/Nile Basin = East African/Nile
Basin Power Pool; SAPP = Southern African Power Pool; WAPP = West African
Power Pool.
a. Island states are Cape Verde, Madagascar, and Mauritius.

Annualized capital investment costs (see box 3.1 for definitions of this
and other cost categories) range from 2.2 percent of the region's GDP
under trade stagnation to 2.4 percent under trade expansion. Regional
annualized capital investment costs under trade expansion exhibit considerable variation: 2 percent of GDP in SAPP, 2.8 percent in WAPP, 3.1 percent in EAPP/Nile Basin, and 1.8 percent in CAPP (table 3.6).

The costs of operating the entire power system are of a similar order
of magnitude. Annualized operating costs range from 1.7 percent of GDP
under trade expansion to 2.1 percent under trade stagnation. The variation among regions under trade expansion is even more pronounced here:
1.7 percent of GDP in SAPP, 2.6 percent in EAPP/Nile Basin, 1.4 percent
in WAPP, and a negligible 0.2 percent in CAPP.

Total annualized costs of system expansion and operation are, therefore,
4.2 percent of GDP under trade expansion and 4.4 percent under trade
stagnation. The regional figures for SAPP and WAPP are similar: 3.7 percent
and 4.2 percent, respectively, under trade expansion, and 3.9 percent and
4.4 percent under trade expansion. Total costs in EAPP/Nile Basin are
higher: 5.7 percent and 6 percent of GDP under trade expansion and
trade stagnation, respectively. They are lower in CAPP: 2 percent under
trade expansion and 2.2 percent under trade stagnation. Around two-
thirds of overall system costs are associated with generation infrastructure
and the remaining one-third with T&D infrastructure.

The overall cost of developing the power system appears high but not
unattainable relative to the GDP of each of the trading regions. Among
countries within each region, however, both GDP and power investment

Table 3.5 Required Spending for the Power Sector in Africa,[a] 2005–15
$ million

Pool	Total expenditure	Total operations and maintenance	Total investment	Investment		
				Rehabilitation	New generation	New T&D
CAPP	1,386	159	1,227	76	860	292
EAPP/Nile Basin	15,004	6,807	8,198	485	5,378	3,334
SAPP	18,401	8,359	10,042	2,554	4,544	2,944
WAPP	12,287	4,049	8,238	1,010	3,527	3,701
Island states[b]	556	311	245	15	74	156
Total	47,634	19,685	27,950	4,140	14,383	10,427

Source: Adapted from Rosnes and Vennemo 2008.

Notes: Assuming national targets for access rates in the trade expansion scenario. CAPP = Central African Power Pool; EAPP/Nile Basin = East African/Nile Basin Power Pool; SAPP = Southern African Power Pool; WAPP = West African Power Pool; T&D = transmission and distribution.

a. Including the Arab Republic of Egypt.

b. Island states are Cape Verde, Madagascar, and Mauritius.

Box 3.1

Definitions

Overnight investment costs. The total cost of expanding the power system to meet demand in 2015. This includes both new investment and refurbishment costs, but not variable costs.

Annualized capital investment costs. The capital investment spending needed each year to meet demand in 2015, taking into account both the discount rate and the varying economic lifetimes of different investments. The formula is as follows:

$$annualized\ capital\ cost = investment\ cost \times r/[1-(1+r)^{-T}],$$

where r is the discount rate (assumed to be 12 percent) and T is the economic lifetime of the power plant (assumed to be 40 years for hydropower plants, 30 years for coal plants, and 25 years for natural gas plants).

The total annualized capital cost refers to both the cost of new generation capacity and the refurbishment of existing capacity, as well as investments in and refurbishment of T&D assets.

Annual variable cost. The costs of fuel and variable costs of operation and maintenance of the system. This includes both existing capacity in 2005 that will still be operational in 2015 and new capacity that will be developed before 2015.

Total annualized cost of system expansion. Annualized capital investment costs plus annual variable costs for new capacity. Variable costs associated with operation of existing capacity in 2005 (generation or transmission) are not included.

Total annualized costs of system expansion and operation. Annualized capital investment costs plus total annual variable costs (for both existing capacity in 2005 and new capacity).

Source: Rosnes and Vennemo 2008.

requirements vary widely. As a result, in certain scenarios some countries face power spending requirements that are very burdensome relative to the size of their economies (figure 3.1). In SAPP, for example, investment requirements exceed 6 percent of GDP in the Democratic Republic of Congo, Mozambique, and Zimbabwe under both trade expansion and stagnation. Spending is similarly high in Egypt, Burundi, and Ethiopia in EAPP/Nile Basin. About half of the countries in WAPP have investment requirements of almost 10 percent of GDP—Guinea and Liberia stand

Table 3.6 Estimated Cost of Meeting Power Needs of Sub-Saharan Africa under Two Trade Scenarios

Scenario	Southern African Power Pool ($billion)	(% GDP)	East African/Nile Basin Power Pool ($billion)	(% GDP)	Western African Power Pool ($billion)	(% GDP)	Central African Power Pool ($billion)	(% GDP)	Total Sub-Saharan Africa ($billion)	(% GDP)
Trade expansion										
Total estimated cost	18.4	3.7	15.0	5.7	12.3	4.2	1.4	2.0	47.6	4.2
Capital costs	10.0	2.0	8.2	3.1	8.2	2.8	1.2	1.8	27.9	2.4
Operating costs	8.4	1.7	6.8	2.6	4.0	1.4	0.2	0.2	19.7	1.7
Generation	11.1	2.2	10.5	4.0	6.5	2.2	1.0	1.4	29.5	2.6
T&D	7.3	1.5	4.5	1.7	5.8	2.0	0.4	0.6	18.1	1.6
Trade stagnation										
Total estimated cost	19.5	3.9	16.0	6.0	12.7	4.4	1.5	2.2	50.3	4.4
Capital costs	10.0	2.0	6.3	2.4	8.0	2.7	1.1	1.6	25.6	2.2
Operating costs	9.4	1.9	9.7	3.7	4.8	1.6	0.4	0.6	24.7	2.2
Generation	12.6	2.5	11.6	4.4	7.1	2.4	1.2	1.7	32.8	2.9
T&D	6.9	1.4	4.4	1.7	5.7	1.9	0.3	0.5	17.5	1.5

Source: Rosnes and Vennemo 2008.

Note: Assumes sufficient expansion to meet national electrification targets. Subtotals may not add to totals because of rounding. GDP = gross domestic product; T&D = transmission and distribution.

Figure 3.1 Overall Power Spending by Country in Each Region

percent GDP

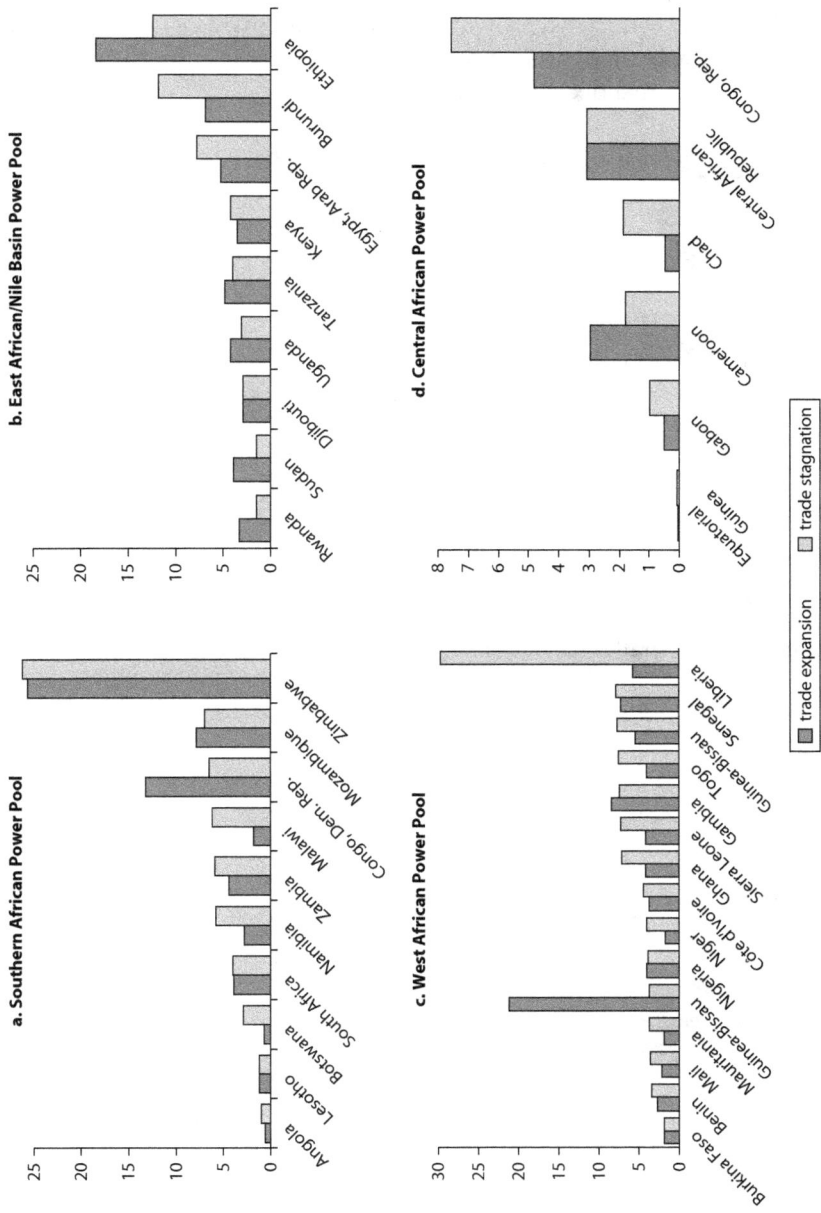

a. Southern African Power Pool

b. East African/Nile Basin Power Pool

c. West African Power Pool

d. Central African Power Pool

■ trade expansion   □ trade stagnation

Source: Rosnes and Vennemo 2008.

out with requirements of almost 30 percent. In CAPP, only the Republic of Congo requires investments of more than 5 percent of GDP.

The next sections explore investment requirements and costs in more detail for each region. More detailed output tables for each country can be found in appendix 3 at the end of this book.

The SAPP

Table 3.7 provides an overview of generation capacity and the capacity mix in SAPP in all scenarios in 2015. The rest of this section provides a description of three trade expansion scenarios.

Constant Access Rates under Trade Expansion

In this scenario, SAPP will require almost 31,300 MW of new capacity to meet demand under trade expansion in 2015. An additional 28,000 MW of existing capacity will need to be refurbished.[4] South Africa accounts for about 80 percent of electricity demand in SAPP. As a result, development there has a strong effect on the rest of the region. Investments in new generation capacity in South Africa amount to

Table 3.7   Generation Capacity and Capacity Mix in SAPP, 2015

	Trade expansion scenario			Trade stagnation scenario	Low-growth scenario
		Regional target access rate	National targets for access rates	National targets for access rates	National targets for access rates, trade expansion
	Constant access rate				
Generation capacity (MW)					
Installed	17,136	17,136	17,136	17,136	17,136
Refurbishment	28,029	28,035	28,046	28,148	28,046
New investments	31,297	32,168	33,319	32,013	20,729
Generation capacity mix (%)					
Hydro	33	33	34	25	40
Coal	60	60	59	66	52
Gas	0	0	0	2	0
Other	7	7	7	7	8

Source: Rosnes and Vennemo 2008.
Note: "Installed capacity" refers to installed capacity as of 2005 that is not refurbished before 2015. Existing capacity that is refurbished before 2015 is included in the definition of "refurbished capacity." SAPP = Southern African Power Pool; MW = megawatt.

18,700 MW (60 percent of the region's total). In addition, 21,700 MW of capacity is refurbished. Coal-fired power plants account for the largest share of capacity investments in South Africa. Open-cycle gas turbine generators[5] account for another 3,000 MW, and hydropower and pumped storage for 2,000 MW.

Elsewhere in SAPP, countries that are rich in hydropower develop substantial new capacity: 7,200 MW in the Democratic Republic of Congo, 3,200 MW in Mozambique, and 2,200 MW in Zimbabwe. In 2005, Zimbabwe imported 14 percent of its electricity, and the new capacity allows the country to meet domestic demand. The Democratic Republic of Congo and Mozambique, on the other hand, export 50 and 6 TWh, respectively, to the rest of the region.

The investment cost of expanding the generation system in SAPP is almost $38 billion (table 3.8). Investments in new capacity account for $30.3 billion, and refurbishment costs account for $7.5 billion. In general, refurbishment is much cheaper than developing new capacity. Therefore,

Table 3.8   Overnight Investment Costs in SAPP, 2005–15
$ million

	Trade expansion scenario			Trade stagnation scenario	Low-growth scenario
	Constant access rate	Regional target access rate	National targets for access rates	National targets for access rates	National targets for access rates, trade expansion
Generation					
Investment cost	30,277	31,103	32,242	34,644	18,589
Refurbishment cost	7,572	7,574	7,577	7,587	7,577
T&D					
Investment cost	16,384	19,422	23,711	20,653	16,606
Cross-border transmission lines	3,009	2,991	3,058	0	3,082
Distribution grid	12,674	12,674	12,674	12,674	5,544
Connection cost (urban)	643	2,210	3,995	3,995	3,995
Connection cost (rural)	58	1,547	3,985	3,985	3,985
Refurbishment cost	9,775	9,775	9,775	9,775	9,775
Total	64,008	67,874	73,304	72,659	52,546

Source: Rosnes and Vennemo 2008.
Note: SAPP = Southern African Power Pool; T&D = transmission and distribution.

despite the large funding gap between the two, refurbishment and new investment make roughly the same contributions (in MW) to new capacity. Coal power plants in South Africa are an exception: Refurbishing them is almost as expensive as investing in new plants.

The additional costs necessary to bring power from power plants to consumers—the costs of T&D and connection—are also substantial: Investments to expand and refurbish the grid total $16 billion (see table 3.8). The direct cost of connecting new customers to the grid is only $0.7 billion, more than 90 percent of which would be spent in urban areas.

Total overnight investment costs are therefore slightly more than $64 billion. Annualized capital costs are $8.8 billion, including $5.6 billion in generation and $3.2 billion in T&D and connection. Annual variable operating costs (including fuel, operation, and maintenance) are $8.3 billion. Operation of new power plants accounts for approximately $2 billion, and operation of existing and refurbished power plants ($3.2 billion) and the grid ($3.1 billion) accounts for the remaining costs. The total annualized cost of system expansion is 2.2 percent of the region's GDP in 2015, and the total annualized cost of system expansion and operation is 3.4 percent.

Costs vary widely among countries. The costs of generation-capacity expansion are particularly high in countries with large hydropower development: 5.8 percent of GDP in the Democratic Republic of Congo, 6.2 percent in Mozambique, and 8.5 percent in Zimbabwe. Grid-related costs (investments, refurbishment, and operation) are high in countries such as Zimbabwe, Zambia, Namibia, and the Democratic Republic of Congo. Finally, although the costs of generation-capacity expansion are only 0.7 percent of GDP in 2015 in South Africa, the annual variable costs of the new coal-fired power plants are 0.6 percent of GDP.

Regional Target for Access Rate: Electricity Access of 35 Percent on Average

Compared with the constant access rate, meeting the average regional target for electricity access (35 percent) requires an additional investment of almost $3.9 billion, or about $0.5 billion in annualized capital costs. The cost of connecting new households accounts for the majority of the additional costs—about $3 billion, or $380 million in annualized costs. Rural areas account for about 40 percent of connection costs, compared with only 10 percent in the constant access rate scenario.

The region also requires additional generation capacity to meet increased demand. Investment costs are $0.8 billion higher ($120 million in annualized costs) compared with the constant access rate scenario. The additional costs of operating the system (variable costs) are much lower— $50 million annually. Overall, the annualized cost of system expansion is 2.3 percent of the region's GDP in 2015. When variable costs of existing capacity are included, the total annualized cost of system expansion and operation rises to 3.6 percent of GDP.

National Targets for Electricity Access

Compared with the constant access rate scenario, meeting national targets requires an additional investment of $9.3 billion, or almost $1.3 billion in annualized costs. The largest contributors to this increase are the costs of T&D and connection. For example, connecting new households to the grid accounts for about $7.3 billion ($0.9 billion in annualized costs) of the additional costs. The additional costs of investment in generation capacity are $2 billion higher ($280 million in annualized costs). Variable costs of operating the system are only $75 million higher each year. The annualized cost of system expansion is 2.4 percent of the region's GDP in 2015. When variable costs of existing capacity are included, the total annualized costs of system expansion and operation rise to 3.7 percent of GDP.

The EAPP/Nile Basin

Table 3.9 provides an overview of generation capacity and the capacity mix in EAPP/Nile Basin in 2015 in all scenarios. The rest of this section provides a description of three trade expansion scenarios.

Constant Access Rates under Trade Expansion

In this scenario, EAPP/Nile Basin will require 23,000 MW of new capacity to accommodate market demand growth in 2015. In addition, more than 1,000 MW of existing capacity must be refurbished. This estimate is based on information about the age of facilities and conditions assembled for this study. Therefore, the need for refurbishment in EAPP/Nile Basin— which is much lower than in SAPP—may have been underestimated.

Egypt imports about 40 percent of its electricity (55 TWh) and accounts for approximately 80 percent of total demand in the EAPP/Nile Basin. As a result, development there is of considerable importance for the rest of the region. Natural gas–fired power plants account for almost 7,000 MW of new capacity in Egypt. Elsewhere in EAPP/Nile Basin,

Table 3.9   Generation Capacity and Capacity Mix in EAPP/Nile Basin, 2015

	Trade expansion scenario			Trade stagnation scenario	Low-growth scenario
	Constant access rate	Regional target access rate	National targets for access rates	National targets for access rates	National targets for access rates, trade expansion
Generation capacity (MW)					
Installed	22,132	22,132	22,132	22,132	22,132
Refurbishment	1,369	1,375	1,375	1,381	1,375
New investments	23,045	24,639	25,637	17,972	23,540
Generation capacity mix (%)					
Hydro	49	47	48	28	48
Coal	2	2	2	3	2
Gas	47	48	49	64	45
Other	2	3	4	5	4

Source: Rosnes and Vennemo 2008.
Note: "Installed capacity" in this table refers to capacity in place in 2005 but not refurbished before 2015. Existing capacity that is refurbished before 2015 is included not in the installed capacity figure, but in the refurbishment figure. Data include Egypt. EAPP/Nile Basin = East African/Nile Basin Power Pool; MW = megawatt.

countries with hydropower resources develop substantial new capacity: 8,150 MW in Ethiopia, 3,700 MW in Sudan, 1,200 MW each in Tanzania and Uganda, and 300 MW in Rwanda. In addition, Kenya and Tanzania invest in some coal-fired power plants, and Ethiopia and Sudan become large net exporters.

To meet projected demand, generation capacity in 2015 must be more than twice the 2005 level. Expanding the generation system over 10 years will cost more than $29 billion (see table 3.10). Investments in new capacity accounts for almost all of this, and refurbishment costs are negligible. The costs of T&D and connection total $11 billion, of which investments in the grid account for $7.5 billion. The cost of connecting new customers is $3 billion, or 40 percent of the total grid investment. Rural areas account for 80 percent of connection costs. Refurbishment of the existing grid requires $3.3 billion.

Total overnight investment costs in EAPP/Nile Basin are $40.2 billion. Annualized capital costs are, therefore, approximately $5.3 billion: $4 billion for generation capacity and $1.3 billion for T&D and connection. The annual variable costs of operating the system amount to $5.84 billion. Operation of new power plants accounts for most of this ($4.39 billion),

Table 3.10   Overnight Investment Costs in the EAPP/Nile Basin, 2015
$ million

	Trade expansion scenario			Trade stagnation scenario	Low-growth scenario
	Constant access rate	Regional target access rate	National targets for access rates	National targets for access rates	National targets for access rates, trade expansion
Generation					
Investment cost	28,913	30,802	32,667	18,621	31,275
Refurbishment cost	396	398	398	399	398
T&D					
Investment cost	7,549	16,430	27,385	26,372	26,301
Cross-border transmission lines	1,320	937	1,013	0	964
Distribution grid	3,072	3,072	3,072	3,072	2,037
Connection cost (urban)	2,484	5,263	5,702	5,702	5,702
Connection cost (rural)	674	7,159	17,599	17,599	17,599
Refurbishment cost	3,342	3,342	3,342	3,342	3,342
Total	40,200	50,973	63,793	48,735	61,317

Source: Rosnes and Vennemo 2008.
Note: Data include Egypt. EAPP/Nile Basin = East African/Nile Basin Power Pool; T&D = transmission and distribution.

and operation of existing and refurbished power plants ($0.69 billion) and the grid ($0.76 billion) account for the rest. The total annualized cost of system expansion is therefore 3.6 percent of the region's GDP in 2015. Adding the variable costs of system operation, the total annualized cost of system expansion and operation is 4.2 percent of GDP.

The cost of system expansion in Egypt—the largest country in the region—is 3.8 percent of its GDP. Capital costs are only 0.9 percent, but because the new capacity is gas fired, fuel costs are 3 percent of GDP. Total annualized costs in Ethiopia are 9.2 percent of its GDP in total—the highest figure in the region. However, investments in generation capacity used for exports account for two-thirds of these costs. Investments in T&D lines and variable costs account for the rest. Costs are particularly low in Burundi and Djibouti—between 1 percent and 2 percent of GDP. In other countries in the region, costs are 2.5–3.5 percent of GDP.

Regional Target for Access Rate: Electricity Access of 35 Percent on Average

Compared with the constant access rate scenario, meeting the international target for electricity access (35 percent on average) requires an additional investment of almost $11 billion, or about $1.3 billion in annualized capital costs. Connecting new households to the grid accounts for the majority of additional costs—$9 billion ($1.1 billion in annualized costs). Rural areas account for 60 percent of the connection costs. The region also requires additional generation capacity to meet increased demand. As a result, investment costs are $2 billion higher ($270 million in annualized costs) than in the constant access rate scenario. Variable costs of operating the system are also $700 million higher annually. Overall, the total annualized cost of system expansion and operation increases to 5 percent of GDP in 2015. Because the costs of operating the existing system are only 0.5 percent of GDP, the total annualized cost of expanding the system amounts to 4.4 percent of GDP.

National Targets for Electricity Access

Meeting national targets requires $24 billion more in investment compared with the constant access rate scenario, or approximately $3 billion in annualized capital costs. The largest contributors to the increase are the costs of T&D and connection. Connecting new households to the grid accounts for $20 billion ($2.4 billion annualized costs) of the additional costs. Rural areas account for 75 percent of connection costs. The additional costs of investment in generation capacity are $3.8 billion ($520 million in annualized costs), and the variable costs of operating the system are $1 billion higher than in the constant access rate scenario. In the national targets scenario, the total annualized cost of system expansion and operation is 5.7 percent of the region's GDP. Excluding the costs of operating the existing system, the total annualized cost of system expansion is 5.1 percent.

WAPP

Table 3.11 provides an overview of generation capacity and the capacity mix in WAPP for all scenarios in the region in 2015. The rest of this section provides a description of three trade expansion scenarios.

Constant Access Rates under Trade Expansion

In this scenario, WAPP requires almost 16,000 MW of new capacity to meet market demand growth in 2015. Almost all of this is hydropower: 10,290 MW in Nigeria, 4,290 MW in Guinea, 1,000 MW in Ghana, and

Table 3.11   Generation Capacity and Capacity Mix in WAPP, 2015

	Trade expansion scenario			Trade stagnation scenario	Low-growth scenario
	Constant access rate	Regional target access rate	National targets for access rates	National targets for access rates	National targets for access rates, trade expansion
Generation capacity (MW)					
Installed	4,096	4,096	4,096	4,096	4,096
Refurbishment	5,530	6,162	6,972	6,842	5,535
New investments	15,979	16,634	18,003	16,239	17,186
Generation capacity mix (%)					
Hydro	82	79	77	73	80
Coal	1	1	1	1	1
Gas	13	14	16	19	12
Other	4	5	6	7	7

Source: Rosnes and Vennemo 2008.
Note: "Installed capacity" refers to installed capacity as of 2005 that is not refurbished before 2015. Existing capacity that is refurbished before 2015 is included in the "refurbished capacity." WAPP = West African Power Pool; MW = megawatt.

130 MW in Côte d'Ivoire. This means that the available hydropower resources become fully exploited[6] in Nigeria, Guinea, and Ghana.[7] One coal-fired power plant (250 MW) is also built in Senegal, and some off-grid technologies are built in rural areas.[8] In addition to investments in new generation capacity, 5,530 MW of existing capacity is refurbished: almost 4,000 MW of hydropower (2,850 in Nigeria), 1,200 MW of natural gas–fired power in Nigeria, and 410 MW of heavy fuel oil (HFO) –fueled thermal power plants in various countries.

Nigeria accounts for two-thirds of electricity consumption in the region. Hence, developments in Nigeria that influence electricity demand (such as economic development and the politically determined electricity access targets) have a large impact on the total cost of electricity sector development in the rest of the region. However, Nigeria does not have a big impact on the trade patterns and resource development in the rest of the region for two reasons. First, Nigeria is not centrally situated and would require large investments in transmission lines to allow for large exports. Second, Nigeria uses its large and relatively cheap hydropower resources to meet domestic demand growth. The ample gas resources that could be used to develop gas-fired power plants are more expensive than hydropower in other countries.

Ghana accounts for 15 percent and Côte d'Ivoire accounts for 6 percent of the region's demand. In contrast with Nigeria, these countries import about half of their electricity. Guinea accounts for almost 20 percent of the region's production and exports more than eight times its domestic demand (mostly competitively priced hydro power).

The investment cost of expanding the generation system in WAPP is slightly more than $23.3 billion (table 3.11). Investments in new capacity account for the majority of this ($22 billion), but the cost of refurbishment is only $1.4 billion.

The costs of T&D and connection are almost equal to the costs of new generation capacity: $23.3 billion for investments to expand and refurbish the grid (table 3.12). Investments in new T&D lines account for more than $17 billion of this. Only 6 percent of this last figure is related to international transmission lines. The direct cost of connecting new customers to the grid is $4.3 billion, or less than 20 percent of the total grid cost. Rural areas account for 86 percent of this total.

Table 3.12 Overnight Investment Costs in WAPP, 2005–15
$ million

| | Trade expansion scenario | | | Trade stagnation scenario | Low-growth scenario |
	Constant access rate	Regional target access rate	National targets for access rates	National targets for access rates	National targets for access rates, trade expansion
Generation					
Investment cost	21,955	23,632	26,992	25,822	25,128
Refurbishment cost	1,363	1,429	1,511	1,496	1,366
T&D					
Investment cost	17,241	22,399	29,813	28,872	23,206
Cross-border transmission lines	1,022	968	941	0	912
Distribution grid	11,909	11,909	11,909	11,909	5,332
Connection cost (urban)	3,698	5,254	7,634	7,634	7,634
Connection cost (rural)	612	4,268	9,329	9,329	9,329
Refurbishment cost	6,057	6,057	6,057	6,057	6,057
Total	46,615	53,518	64,373	62,247	55,758

Source: Rosnes and Vennemo 2008.
Note: WAPP = West African Power Pool; T&D = transmission and distribution.

Total overnight investment costs are $46.6 billion in this scenario. The annualized capital cost of meeting market demand in 2015 is $6 billion: almost $3 billion in T&D and connection and $3.1 billion in generation. The annual variable operating costs are $3.2 billion. About half of this is related to operating new power plants, and the other half is related to operating existing and refurbished power plants ($0.3 billion) and the grid ($1.3 billion). The total annualized cost of system expansion is therefore equivalent to 2.1 percent of the region's GDP in 2015. Adding the variable operation costs of existing capacity, the total annualized cost of system expansion and operation is 3.2 percent of GDP.

Investment patterns, and therefore costs, vary widely among countries in the region. For example, Guinea invests in hydropower for export purposes, and the total investment costs are 20 percent of GDP. In The Gambia, variable fuel costs of existing HFO-fueled capacity are 4.5 percent of GDP, and the grid cost makes up another 1 percent of GDP. In Senegal, both the grid-related cost (investment and variable) and variable generation cost contribute to raising the total cost to 7 percent of GDP.

Regional Target Rate: Electricity Access of 54 Percent on Average

Compared with the constant access rate scenario, meeting the regional target for electricity access (54 percent on average) requires additional investment of almost $7 billion, or about $1.25 billion in annualized capital costs. Connecting new households to the grid accounts for the majority of additional costs—more than $5 billion ($600 million in annualized costs). Almost half of this amount is spent in rural areas. The region also requires additional generation capacity to meet increased demand: Investment costs are $1.7 billion higher ($200 million in annualized costs) than in the constant access rate scenario. Variable operating costs are 12 percent higher (almost $400 million annually) because part of the new generation capacity is supplied by fossil fuels (diesel in rural areas and refurbishment of gas-fired power plants in Nigeria). The total annualized cost of system expansion is 2.4 percent of the region's GDP in 2015. Including variable costs of existing capacity lifts the total annualized cost of system expansion and operation to 3.6 percent of GDP.

National Targets for Electricity Access

Compared with the constant access rate scenario, meeting national targets requires an additional investment of $18 billion, or approximately $3 billion in annualized costs. The largest contributors to this increase are the costs of T&D and connection. For example, connecting new households to the grid involves an extra investment of $12.5 billion ($1.6 billion in annualized costs). Rural areas account for more than half of connections. The region also requires additional investment in generation capacity to meet increased demand: Investment costs are more than $5 billion higher ($650 million in annualized costs). The variable operating costs are $850 million annually. The total annualized cost of system expansion is 2.9 percent of the region's GDP in 2015. When variable costs of existing capacity are included, the total annualized costs of system expansion and operation rise to 4.2 percent of GDP.

CAPP

Table 3.13 provides an overview of generation capacity and the capacity mix in CAPP for all scenarios. The rest of this section provides a description of three trade expansion scenarios.

Constant Access Rates under Trade Expansion
CAPP requires 3,856 MW of new capacity to meet market demand growth in 2015. All of this is hydropower:[9] 2,430 MW in Cameroon, 1,318 MW in the Republic of Congo, 84 MW in Gabon, and 24 MW in the Central African Republic. This means that the available hydropower

Table 3.13   Generation Capacity and Capacity Mix in CAPP, 2015

	Trade expansion scenario			Trade stagnation scenario	Low-growth scenario
	Constant access rate	Regional target access rate	National targets for access rates	National targets for access rates	National targets for access rates, trade expansion
Generation capacity (MW)					
Installed	260	260	260	260	260
Refurbishment	906	906	906	1,081	906
New investments	3,856	4,143	4,395	3,833	3,915
Generation capacity mix (%)					
Hydro	97	97	97	83	97
Coal	0	0	0	0	0
Gas	0	0	0	0	0
Other	2	3	3	17	3

Source: Rosnes and Vennemo 2008.
Note: "Installed capacity" refers to installed capacity as of 2005 that is not refurbished before 2015. Existing capacity that is refurbished before 2015 is included in the "refurbished capacity." CAPP = Central African Power Pool; MW = megawatt.

resources are fully exploited in Cameroon.[10] In addition, more than 900 MW of existing capacity must be refurbished. Cameroon accounts for 600 MW of refurbished capacity, and Gabon, the Republic of Congo, and the Central African Republic account for the rest.

The Republic of Congo accounts for more than one-half (54 percent) of electricity demand in CAPP in 2015, and Cameroon accounts for one-third. Therefore, the development of these two countries has a strong effect on the rest of the region. Gabon has 10 percent of the region's total demand, but the other countries have minimal electricity demand.

Cameroon accounts for 64 percent of total electricity production in the region in 2015, and the Republic of Congo accounts for only 29 percent. Cameroon exports more than one-third of its production (5.6 TWh) to the Republic of Congo and exports small amounts to Gabon, Chad, and Equatorial Guinea. It is assumed that imports from the Democratic Republic of Congo to the Republic of Congo remain at their 2005 levels, but this is a small volume (less than 0.5 TWh per year).

The investment cost of expanding the generation system in CAPP is almost $6 billion (table 3.14). Investments in new capacity account for

Table 3.14   Overnight Investment Costs in CAPP, 2005–15
$ million

	Trade expansion scenario			Trade stagnation scenario	Low-growth scenario
	Constant access rate	Regional target access rate	National targets for access rates	National targets for access rates	National targets for access rates, trade expansion
Generation					
Investment cost	5,645	6,157	6,615	5,981	5,766
Refurbishment cost	272	272	272	301	272
T&D					
Investment cost	1,057	1,648	2,348	2,036	2,311
Cross-border	349	317	312	0	355
Distribution grid	286	286	286	286	205
Connection cost (urban)	412	753	1,010	1,010	1,010
Connection cost (rural)	10	292	740	740	740
Refurbishment cost	222	222	222	222	222
Total	7,196	8,299	9,457	8,540	8,570

Source: Rosnes and Vennemo 2008.
Note: CAPP = Central African Power Pool; T&D = transmission and distribution.

the majority of this ($5.6 billion), while the cost of refurbishment is only $0.3 billion. The costs of T&D and connection are much lower than the costs of building new power plants and account for less than 20 percent of total investment costs. The costs of expanding and refurbishing the grid are $1.3 billion, most of which (over $1 billion) is investment in new T&D lines. A third of this last figure is related to international transmission lines. The direct cost of connecting new customers to the grid is 40 percent of the total grid investment cost, or $0.4 billion. Urban areas account for 98 percent of connection costs.

Total overnight investment costs in CAPP in the constant access rate scenario are slightly more than $7 billion. The annualized capital cost of meeting market demand through 2015 is therefore almost $1 billion: $780 million in generation and almost $160 million in T&D and connection. Annual variable operating costs amount to $150 million, about $50 million of which is related to operating new power plants. The rest is related to operating existing and refurbished power plants ($30 million) and the grid ($70 million). The total annualized cost of system expansion is about $1 billion, equivalent to 1.4 percent of the region's GDP in 2015. Adding the variable operation costs of existing capacity, the total annualized cost of system expansion and operation is 1.6 percent of GDP.

Investment patterns and costs vary widely among countries in the region. In particular, the costs of expanding generation are high in countries with relatively large hydropower development: 3 percent of GDP in the Republic of Congo and 1.6 percent in Cameroon. The Republic of Congo also imports a substantial amount of electricity. Grid-related costs (including investments, refurbishment, and operation) account for another 0.3 percent of GDP in the Republic of Congo, mainly because new cross-border lines need to be built to make the large imports possible. Grid-related costs are 0.3 percent of GDP in Cameroon as well. This is mainly due to connecting new customers to the grid, in addition to investments in the domestic and cross-border grids. Finally, Chad and Equatorial Guinea do not invest in any new generation capacity. Their costs are related to grid expansion, maintenance, and new connection, which are relatively inexpensive.

Regional Target for Access Rate: Electricity Access of 44 Percent on Average

Compared with the constant access rate scenario, meeting the international target for electricity access (44 percent on average) requires an additional investment of $1.1 billion, or about $140 million in annualized capital costs. Connecting new households to the grid accounts for about

$0.6 billion ($80 million annually) of total additional costs. Almost 30 percent of the total connection costs are spent in rural areas, compared with only 2 percent in the constant access rate scenario. The region also requires additional generation capacity to meet increased demand: Investment costs are $0.5 billion higher ($66 million in annualized costs) than in the constant access rate scenario. Variable operating costs are slightly higher because some of the new generation capacity in rural areas is based on off-grid diesel generators (there is also some mini-hydro and solar photovoltaic in the rural areas). The total annualized cost of system expansion is therefore 1.6 percent of the region's GDP in 2015. Including variable costs of existing capacity lifts the total annualized cost of system expansion and operation to 1.8 percent of GDP.

National Targets for Electricity Access

Meeting national targets requires $2.3 billion more in investment than keeping the access rate constant at current levels. This corresponds to $300 million in annualized costs. The largest contributors to this increase are the costs of T&D and connection. For example, connecting new households to the grid involves an extra cost of about $1.3 billion ($165 million in annualized costs). More than 40 percent of the total costs of new connections are spent in rural areas, compared with only 2 percent in the constant access rate scenario. The region also requires additional generation capacity to meet demand: Investment costs are almost $1 billion higher ($126 million in annualized costs). The total annualized cost of system expansion is therefore 1.8 percent of GDP in 2015. Including the variable operating costs of existing capacity increases the total annualized cost of system expansion and operation to 2 percent of GDP.

Notes

1. Data for Sub-Saharan Africa exclude Egypt.

2. The membership of the power pool is as follows: SAPP: Angola, Botswana, the Democratic Republic of Congo, Lesotho, Malawi, Mozambique, Namibia, South Africa, Zambia, and Zimbabwe. EAPP: Burundi, Djibouti, Egypt, Ethiopia, Kenya, Rwanda, Sudan, Tanzania, and Uganda. WAPP: Benin, Burkina Faso, Côte d'Ivoire, The Gambia, Ghana, Guinea, Guinea-Bissau, Liberia, Mali, Mauritania, Niger, Nigeria, Senegal, Sierra Leone, and Togo. CAPP: Cameroon, the Central African Republic, Chad, the Republic of Congo, Equatorial Guinea, and Gabon.

3. Notional demand refers to the aggregate quantity of goods and services that would be demanded if all markets were in equilibrium.

4. This includes both power plants that were operational in 2005 but will need to be refurbished before 2015 and plants that were not operational in 2005.

5. South Africa has already committed to building 3,000 MW of capacity in open-cycle gas turbine generators. This capacity is therefore included exogenously in the model.

6. Fully exploited refers to the assumed maximum potential for hydropower in the model. In most cases, this maximum potential has been set equal to identified projects and plans, even though the full hydropower potential of a country may be much larger. The identified projects serve as a proxy for developments that are realistic in the time frame in focus here (the next 10 years, formally before 2015).

7. Because we use only one (average) investment cost per technology per country, not individual costs per project, cheaper resources are often fully utilized in one country before the more expensive resources are developed in a neighboring country. The cost of building international transmission lines counteracts this to some extent.

8. In addition, there are tiny investments in off-grid technologies in rural areas.

9. There are negligible investments in off-grid technologies in rural areas.

10. See note 6 for a definition of "fully exploited."

Reference

Rosnes, Orvika, and Haakon Vennemo. 2008. "Powering Up: Costing Power Infrastructure Spending Needs in Sub-Saharan Africa." Background Paper 5, Africa Infrastructure Country Diagnostic, World Bank, Washington, DC.

CHAPTER 4

Strengthening Sector Reform and Planning

Since the 1990s reform has swept across the power sector in developing regions. Sub-Saharan Africa is no exception. New electricity acts have been adopted that envisage the reform of state-owned electricity utilities and permit private sector participation. Thus far, however, the private sector has had only limited involvement in reforms. Various short-term private management contracts were awarded, but few have resulted in sustainable improvements in the performance of national utilities. Only a few private leases and concessions survive, mostly in Francophone West Africa. The private sector has been involved primarily in the generation sector.

Sub-Saharan Africa's deficit in generation capacity and lack of investment resources has opened the door for independent power projects (IPPs). Power sector reforms originally followed the prescription of industry unbundling, privatization, and competition, but electricity markets that meet these criteria are nowhere to be found in Africa. Instead, the region has seen the emergence of hybrid markets in which incumbent state-owned utilities often retain dominant market positions and IPPs are introduced on the margin of the sector.

Attracting investment to hybrid power markets presents new challenges. Confusion arises about who holds responsibility for power sector planning, how procurement should be managed, and how to allocate investment among state-owned utilities and IPPs. These challenges need to be addressed if the generation sector in Sub-Saharan Africa is to benefit from the promised new private investment.

Independent electricity or energy regulatory agencies have also been established in most Sub-Saharan African countries. They were originally intended to protect consumers, facilitate market entry, and provide price certainty for investors, but they are now criticized for inconsistent decision making and for exacerbating regulatory risk. Independent regulation depends on adequate political commitment and competent, experienced institutions. Without these prerequisites, other forms of regulation may be preferable, such as those that curtail regulatory decision-making discretion with more specific legislation, rule, and contracts. Some regulatory functions may also be outsourced to expert panels.

Power Sector Reform in Sub-Saharan Africa

Power sector reform in Sub-Saharan Africa has been widespread. There have been attempts to improve the performance of state-owned utilities, new regulatory agencies have been created, private management contracts and concessions have been awarded, and private investment has been sought in the form of IPPs.

As of 2006, all but a few of the 24 countries of Sub-Saharan Africa covered by the Africa Infrastructure Country Diagnostic (AICD) had enacted a power sector reform law, three-quarters had introduced some form of private participation, two-thirds had privatized their state-owned power utilities, two-thirds had established a regulatory oversight body, and more than one-third had independent power producers (figure 4.1). About one-third of the countries have adopted three or four of those reform components, but few have adopted all of them, and the extent of reform remains limited. In most countries, for example, the national state-owned utility retains its dominant market position. Private sector cooperation is either temporary (for example, a limited-period management contract) or marginal (in the form of independent power producers that contract with the state-owned national utility). In most cases, the national utility is the mandated buyer of privately produced electricity while still maintaining its own generation plants. There is no wholesale or retail competition in Africa.[1]

Many countries are reconsidering whether certain reform principles and programs—notably the unbundling of the incumbent utility to foster

Figure 4.1   Prevalence of Power Sector Reform in 24 AICD Countries

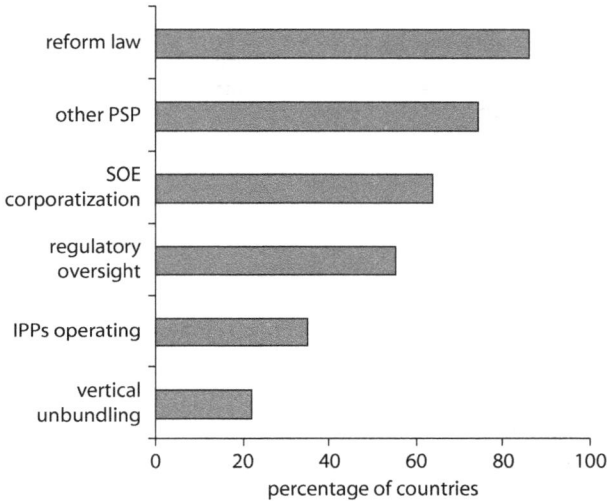

Source: Eberhard 2007.
Note: "Other PSP" means forms of private sector participation other than independent power projects (IPPs), namely, concessions or management contracts. AICD = Africa Infrastructure Country Diagnostic; SOE = state-owned enterprise.

competition—are appropriate for Sub-Saharan Africa.[2] Besant-Jones (2006), in his global review of power sector reform, concludes that power sector restructuring to promote competition should be limited to countries large enough to support multiple generators operating at an efficient scale, which excludes most countries of Sub-Saharan Africa. Even South Africa and Nigeria, which are large enough to support unbundling, have not seen much progress.

An examination of the database on private participation in infrastructure (PPI) maintained by the Public-Private Infrastructure Advisory Facility (PPIAF), which covers all countries in Sub-Saharan Africa, unearthed nearly 60 medium- to long-term power sector transactions involving the private sector in the region (excluding leases for emergency power generation). Almost half are IPPs, accounting for nearly 3,000 megawatts (MW) of new capacity and involving more than $2 billion of private sector investment (table 4.1). Côte d'Ivoire, Ghana, Kenya, Mauritius, Nigeria, Tanzania, and Uganda each support two or more IPPs. A few IPP investments have been particularly successful, including the

Table 4.1   Overview of Public-Private Transactions in the Power Sector in Sub-Saharan Africa

Type of private participation	Countries affected	Number of transactions	Number of canceled transactions	Investment in facilities ($ million)
Management or lease contract	Chad, Gabon, Gambia, Ghana, Guinea-Bissau, Kenya, Lesotho, Madagascar, Malawi, Mali, Namibia, Rwanda, São Tomé and Príncipe, Tanzania, Togo	17	4	5
Concession contract	Cameroon, Comoros, Côte d'Ivoire, Gabon, Guinea, Mali, Mozambique, Nigeria, Sao Tomé and Príncipe, Senegal, South Africa, Togo, Uganda	16	5	1,598
Independent power project	Angola, Burkina Faso, Republic of Congo, Côte d'Ivoire, Ethiopia, Ghana, Kenya, Mauritius, Nigeria, Senegal, Tanzania, Togo, Uganda	34	2	2,457
Divestiture	Cape Verde, Kenya, South Africa, Zambia, Zimbabwe	7	—	n.a.
Overall		74	11	4,060

Source: World Bank 2007; AICD 2008.
Note: — = data not available; n.a. = not applicable.

Tsavo IPP in Kenya (box 4.1) and the Azito power plant in Côte d'Ivoire (box 4.2).

Gratwick and Eberhard (2008) predict that although IPPs have sometimes been costly because of technology choices, procurement problems, and currency devaluation, they will nevertheless continue to expand generation capacity on the continent. Some have been subject to renegotiation. Several factors contribute to the success of IPPs: policy reforms, a competent and experienced regulator, timely and competitive bidding and procurement processes, good transaction advice, a financially viable off-taker, a solid power-purchase agreement (PPA), appropriate credit and security arrangements, availability of low-cost and competitively priced fuel, and development-minded project sponsors.

The other half of the PPI transactions in Sub-Saharan Africa have been concessions, leases, or management contracts, typically for the operation of the entire national power system. Many of these projects have

Box 4.1

Kenya's Success with Private Sector Participation in Power

Private sector participation in the power sector in Kenya started with the Electric Power Act of 1997. Since then Kenya has implemented important electricity reforms. The act also introduced independent economic regulation in the sector, which is important for creating a more predictable investment climate to encourage public sector participation. It has since become government policy that all bids for generation facilities are open to competition from both public and private firms and that the national generator does not receive preferential treatment.

The sector was unbundled in 1998 with the establishment of the Kenya Electricity Generating Company (KenGen, generation) and Kenya Power and Lighting Company (KPLC, transmission and distribution). Now KenGen and KPLC are 30 percent and 50 percent privately owned, respectively.

The Electricity Regulatory Board was established in 1998. It was converted into the Energy Regulatory Commission and granted new powers in 2007. To date the government has not overturned a decision of the board or commission, and it maintains a significant degree of autonomy. It has issued rules on complaints and disputes, licenses, and tariff policy. The regulator also oversees generation expansion planning. KPLC manages the procurement and contracting process with IPPs, subject to approval by the regulator of power purchase agreements.

Five independent power producers supply an increasing proportion of the country's electricity, and three additional IPPs have recently been bid out. A proposed wind farm has also recently been licensed (but not yet built). An independent evaluation by the University of Cape Town (Gratwick and Eberhard 2008) concluded that IPPs had a positive outcome on the development of Kenya's power sector. The public sector developed very little generation capacity in the decade preceding reforms. The performance of KenGen's existing plant is inferior to adjacent IPPs. The Tsavo IPP in Kenya is a particularly good example of an investment that came through an international competitive bidding process and subsequently produced reliable and competitively priced power.

Source: Authors' compilation based on background materials provided by the World Bank's Africa Energy Department staff, 2009.

Box 4.2

Côte d'Ivoire's Independent Power Projects Survive Civil War

Compagnie Ivoirienne de Production d'Electricité (CIPREL) was among the first IPPs in Africa. CIPREL began producing power in 1994 with a 210 MW open-cycle plant fired by domestically produced natural gas. SAUR Group and Electricité de France (EDF) were major shareholders.

At the time, Côte d'Ivoire's investment climate was among the best in the region, and the economy was growing at an annual rate of 7.7 percent. This favorable climate, coupled with CIPREL's success, stimulated interest in the second IPP, Azito, during its international competitive bid in 1996. Ultimately a consortium headed by Cinergy and Asea Brown Boveri was selected to develop the plant, and the deal was safeguarded by a sovereign guarantee and a partial risk guarantee from the World Bank. In 2000 Azito's 330 MW gas-fired, open-cycle plant came online, becoming the largest IPP in West Africa.

Just months after Azito's deal was finalized and well before the plant was completed, Côte d'Ivoire suffered a political coup. During the years of civil unrest between 1999 and 2007, the revenues of the national utility, Compagnie Ivoirienne d'Electricité (CIE), declined by approximately 15 percent, reducing the state's ability to invest in much-needed electricity infrastructure. Yet the turmoil had no impact on the IPPs, and they continued to produce electricity and make payments to CIE. Both IPPs are keen to expand their interest in the generation sector.

Why have IPPs in Côte d'Ivoire fared so well? A stable currency pegged to the euro (and earlier to the French franc) minimizes the exchange-rate risks that have taxed other Sub-Saharan African IPPs. Cohesive power sector planning after the droughts of the 1980s helped the country achieve a good mix of hydro and thermal power sources. The country has a sufficient power supply for itself and for exports to its neighbors in their times of need. The political instability was also confined to the north of the country, where there are fewer consumers than in the south. This allowed the utility to collect sufficient revenues, even when they stopped flowing in from rebel-controlled areas. The availability of domestic gas also helped keep power prices down. The sponsors of the IPP (SAUR and EDF) have been involved throughout the power supply chain, which may explain why there have been no disruptions and why interest continues. Development partners (the World Bank via the International Development Association and the International Finance Corporation; the West African Bank for Development; Promotion et Participation pour la Cooperation Economique; and firms with a development mandate, such as IPS and Globeleq have played a critical role in finalizing and sustaining the deals.

Source: Gratwick and Eberhard 2008.

been unsuccessful; about one-third of the contracts are either in distress or have already been canceled. Long-term private leases or concessions have survived only in Cameroon, Cape Verde, Côte d'Ivoire, Gabon, Mali, and Uganda.

Private Management Contracts: Winning the Battle, Losing the War

The only remaining private management contracts in the power sector in Sub-Saharan Africa are in Madagascar and The Gambia. After the expiration of management contracts in several other countries (including Namibia, Lesotho, Kenya, Malawi, Tanzania, and Rwanda), utilities reverted to state operation.[3]

Management contracts were once regarded as the entry point for PPI. Because the state retained full ownership of the assets, the government could avoid the political objections that inevitably accompany divestiture. Furthermore, because the private management contractor would neither acquire equity nor incur commercial risk, it should be simple for governments to hire competent professionals, pay them a fee for their services (plus bonuses for fulfillment of specified performance targets in most cases), and enjoy the resulting financial and operational improvements.

In reality, management contracts have proved complex and contentious. Although widely used (there are 17 contracts in 15 countries in the region) and usually productive in terms of improving utility collection rates and revenues and reducing system losses, management contracts have not been able to overcome the broader policy and institutional deficiencies of the sector. Moreover, they have failed to generate much-needed investment funds, either through generating sufficient revenue or through improving investment ratings and attracting private debt. Nor have they proven sustainable. Of the 17 African management contracts, four were canceled before their expiration date, and at least five more were allowed to expire after their initial term (in Gabon and Mali management contracts were followed by concessions).

Why has it proven difficult to implement and retain support for an ostensibly simple management contract? The disconnect among stakeholder expectations bears a large part of the responsibility. Donors and development finance institutions, which have been involved in almost all management contracts, regard it as a first step toward greater liberalization and privatization of the utility and not an end in itself. Yet only in Gabon and Mali did management contracts mark the beginning of further liberalization. Even in countries where concessions or divestitures were clearly not an option (mostly because of popular or political opposition to privatization), donors viewed the contracts as part of a larger reform process and expected them to be extended long enough to allow parallel policy and institutional changes to take root. African governments, on the other hand, saw them not as easy first steps but as undesirable obligations that they needed to fulfill to receive crucial donor funding.

Assessments of the impact of African electricity management contracts indicate improved performance, including greater labor productivity, better collection rates, and reduced system losses. For example, between mid-2002 and mid-2005 under the management contract in Tanzania, collection rates rose from 67 to 93 percent, system losses fell by 5 percent, 30,000 new connections were installed (at a pace far greater than the previous expansion rate), costs fell by 30 percent, and annual revenues rose by 35 percent. Labor relations improved despite the layoff of more than 1,300 workers, whose departure was eased by a generous severance package. The utility introduced a poverty tariff for consumers using 50 kilowatt-hours a month or less (Ghanadan and Eberhard 2007). Working capital overdrafts were cleared, and the utility even secured small loans from private commercial banks (contingent on the continued presence of the management contractors). A management contractor in the rural, northern part of Namibia also produced significant gains. Between 1996 and 2002, the number of customers doubled, and labor productivity soared without a change in the size of the workforce (Clark and others 2005).

Based on the promising results from these and other management contracts, donors concluded that they were an effective method for improving utility performance. Some country officials, however, were more skeptical. They acknowledged that performance had improved but argued that they were largely a result of foreign managers being allowed to lay off excess staff, cut service to delinquent customers, and raise tariffs—African managers in state-owned utilities had not had the same freedom. The main counterargument was therefore that if public managers were given the same authority as management contractors, they could achieve similar performance at a much lower price.

Management contracts may have proved easier to sustain had they been accompanied or followed by large amounts of external investment funding, or had they substantially improved service quality or reduced costs enough to provide investment capital from retained earnings for network rehabilitation and expansion. They were not able to do so, however, partly because of poor initial conditions and partly because they often coincided with cost-raising factors beyond the control of utility managers such as regional drought, soaring oil prices, and the need to purchase expensive power from IPPs.

African ministries of finance were doubtless pleased with the financial and efficiency gains observed under the management contracts. Yet most customers were unaware of or indifferent to financial improvements and were instead concerned with service quantity, quality, and price. In these areas, changes were gradual and modest. Critics of privatization and private participation—including some who had been displaced from management posts by the management contracts—objected to continued load shedding and the indignity of relying on foreign managers. They also protested the substantial contractor fees. For example, the management contractor in Tanzania earned $8.5 million in fixed fees and $8.9 million in performance-based fees during its 56 months in operation. (Those fees were a small fraction of the financial gains produced under the management contract, and the Swedish donor, the Swedish International Development Cooperation Agency, paid a large portion of the performance-based reward). The significant political backlash convinced policy makers that the benefits of management contracts did not outweigh the costs, and the contracts were allowed to lapse.

Although management contracts can improve the efficiency and sustainability of utilities, they cannot overcome the obstacles posed by broader policy and institutional weaknesses. Moreover, the performance improvements are gradually distributed to unaware and unorganized consumers, whereas the costs immediately affect a vocal and organized few, whose protests often overcome rational debate. African management contracts appear to have won the economic battles but lost the political war. They must therefore be restructured to be sustainable and more widely palatable.

Sector Reform, Sector Performance

Sub-Saharan Africa lags behind other regions in installed capacity, electricity production, access rates, costs, and reliability of supply. Many other performance indicators are also subpar. For example, the utilities have an average of only about 150 customers per employee, compared with an average of more than 500 in the high-income member countries of the Organisation for Economic Co-operation and Development. Transmission and distribution (T&D) losses average 25 percent. Commercial efficiency, collection rates, and cost recovery are also poor.

Figure 4.2   Effect of Management Contracts on Performance in the Power Sector in Sub-Saharan Africa[a]

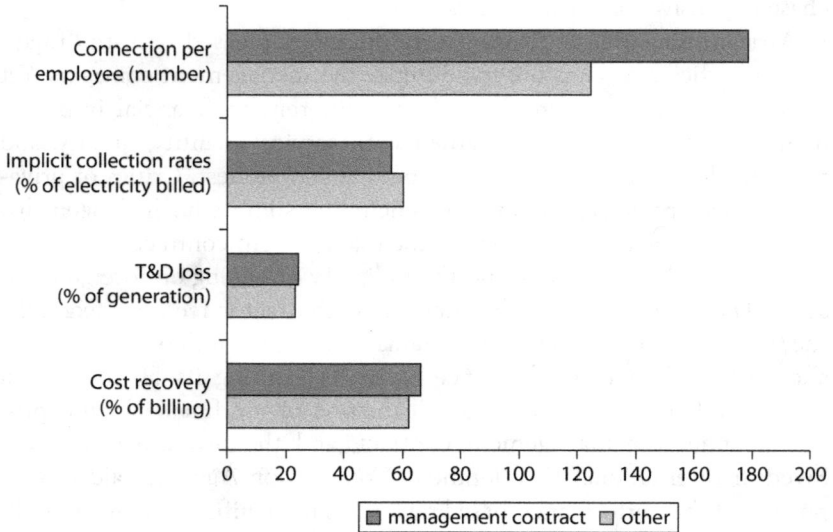

Source: Vagliasindi and Nellis 2010.
Note: T&D = transmission and distribution.
a. Performance differential is statistically significant at the 1 percent level.

Power sector reform should improve utility performance (Gboney 2009). Nevertheless, although PPI generally has a positive effect on performance, it does not always improve all performance indicators (figure 4.2). Disaggregated data on PPI, however, reveal that utilities in countries with IPPs almost always fare better and that concessions are far more effective than management contracts in improving perform-ance. Countries with management contracts fail to make any major or sustained improvements (except in labor productivity).

The Search for Effective Hybrid Markets

The 1990s reform prescription of utility unbundling and privatization followed by wholesale and retail competition was not effective in Africa. Most of the region's power systems are too small to support meaningful competition. The new reality is therefore one of hybrid power markets. In this model the state-owned utility remains intact and occupies a dom-inant market position, whereas private sector participation (typically in the form of IPPs) compensates for the lack of investment on the part of governments and utilities. Africa's hybrid electricity markets pose new challenges in policy, regulation, planning, and procurement, which are compounded by widespread power shortages and an increasing reliance on emergency power throughout the region.

It is often uncertain where responsibility for ensuring adequate and reliable supply lies in hybrid power markets. Few countries in Africa have an explicit security of supply standard,[4] and the incumbent state-owned national utility has typically assumed the responsibility as supplier of last resort. However, few government departments or regulators explicitly monitor adequacy and reliability of supply, and even fewer require utilities to regularly disclose public reports regarding their security of supply. If monitoring were institutionalized, then regulators would be in a better position to assess the need for investment in new capacity.

Traditionally the state-owned utility bore responsibility for planning and procurement of new power infrastructure. With the advent of power sector reforms and the introduction of IPPs, those functions were often moved to the ministry of energy or electricity. A simultaneous transfer of skills did not always occur, however, resulting in poorly executed plans; in many cases generation expansion planning has collapsed.

Where still present, planning tends to take the form of outdated, rigid master plans that do not reflect the changes in price and availability of fuel and equipment and the resulting least-cost options. Planning needs to be dynamic and flexible, and potential investors should benefit from regular disclosure of information regarding demand growth and investment opportunities. At the same time, planning should not preclude the emergence of innovative solutions from the market.

The allocation of responsibility for capacity expansion should be carefully considered. The national utility generally has much greater access to resources and professional staff than either the energy ministry or the regulator. It therefore may be the most pragmatic choice to be the authority for national planning, especially if the transmission and system operations are unbundled from generation. If this is the case, however, a governance and oversight mechanism would be needed to ensure that national interests, and not the interests of the utility, motivate planning. Box 4.3 explores South Africa's difficulties with planning in the power sector.

Incumbent state-owned utilities often argue that they are able to supply power more cheaply or quickly than private alternatives (even if they lack the resources to do so). Yet rigorous analysis that assigns appropriate costs to capital seldom supports such claims, which undermine the entry

Box 4.3

Power Sector Planning Dilemmas in South Africa

The state-owned national utility Eskom dominates South Africa's power market. It generates 96 percent of the country's electricity and through 2006 has provided reliable and secure power supplies. This was largely possible because massive overinvestment in the 1970s and 1980s generated substantial spare capacity. In 1998 the government published a white paper on energy policy, which proposed that Eskom be unbundled, 30 percent be sold, and competition introduced. From 2001 to 2004 consultants worked to design a power exchange and bilateral power market with associated financial contracts for differences, futures, and forward options—not unlike NordPool in Scandinavia or PJM on the East Coast of the United States. During this time the government prohibited Eskom from investing in new capacity because the market would provide new private investment.

Eskom was traditionally the supplier of last resort in South Africa and had responsibility for power sector planning and new investments. Now confusion arose as to who was responsible for these functions. Eskom continued to develop plans, but so did the Ministry of Energy and the regulator—and each differed from the other. At the same time, growing demand and a lack of new capacity were eroding reserve margins. The consultants' plan was never implemented. No new private investment was possible in this context of market uncertainty and in the absence of clear contracting frameworks.

In 2004 the government abandoned its plans to establish a power exchange, and Eskom once again assumed responsibility for expanding generation capacity. At the same time, IPPs would be allowed to enter the market. By this point, Eskom was four years behind in its investments. It has since ordered new large-base-load power stations, but these will begin to come online in 2012. In the meantime, South Africa has experienced power rationing and blackouts.

The government has reassigned responsibility for power sector planning to Eskom, although the Ministry of Energy decides which of Eskom's planning scenarios to adopt. The Ministry then promulgates and publishes the official plan, on which the regulator bases its licensing of generators.

Although this arrangement has provided some certainty regarding the allocation of responsibilities for planning, the official plan is prescriptive rather than indicative and potentially excludes many innovative investment solutions from the private sector. So far no new IPPs have been contracted, although some cogeneration contracts have been concluded. The Ministry of Energy is also

(continued next page)

Box 4.3 *(continued)*

developing a proposal to unbundle the planning, buying, transmission, and system operation functions from Eskom.

The case of South Africa illustrates the complexity and difficulty of involving both state-owned utilities and IPPs in hybrid power markets. In particular, it highlights the importance of clearly allocating responsibility for planning and procurement functions, developing flexible and up-to-date plans, and establishing governance mechanisms to ensure that decisions on capacity expansion and procurement are made transparently, fairly, and in the national interest.

Source: Authors.

of IPPs. Regardless, most African utilities have not supplied adequate investment in much-needed generation capacity.

Poor understanding of the hybrid market prevents policy makers from devising clear and transparent criteria for allocating new building opportunities among the incumbent state-owned utility and IPPs. The failure to order new plants on a timely basis discourages investors and results in power shortages that prompt recourse to expensive emergency power. This has been the case in Tanzania and Rwanda. When authorities finally begin procurement, they may not take the trouble to conduct international competitive bidding. This is unfortunate, because a rigorous bidding process provides credibility and transparency and results in more competitively priced power.

Unsolicited bids can lead to expensive power. The best example of that is IPTL in Tanzania, which provides some of the most costly power in the region (when it is operational, because an unresolved arbitration process has recently closed the plant). However, unsolicited bids sometimes allow private investors to offer innovative generation alternatives, and they generally cover the project development costs. Theoretically, unsolicited bids could be subjected to a Swiss challenge whereby the project is bid out competitively, and the original project developer can subsequently improve their offer to beat the most competitive bid. In practice, however, the Swiss challenge would be difficult to implement if the project developer owns associated fuel resources (for example, a coal field) or if the project is unique is some way (for example, the development of methane resources in Lake Kivu in Rwanda). Governments should therefore opt for international competitive bids when feasible but should also develop policies for handling unsolicited bids.

Hybrid markets also require clarity on the IPP off-take arrangements. For various reasons, power from IPPs in Sub-Saharan Africa is likely to be more expensive than from the national utility. For example, the generation plant for the national utility may be largely depreciated and paid for (for instance, old hydroelectric facilities), and prices may not necessarily reflect costs. Customers are thus likely to seek their power from the state-owned utility rather than buying directly from the IPP (unless security of supply concerns make power from IPPs more attractive, despite higher prices). In most cases, however, IPPs will require off-take agreements with incumbent national utilities that aggregate demand and average prices for customers. Surprisingly few African countries have explicitly defined their power market structures or procedures for negotiating and contracting PPAs with IPPs. Some countries have used the single-buyer model with the national utility as the buyer. Yet it is not always clear whether this implies that the national utility has exclusive purchasing rights. For example, are IPPs required to sell only to the national utility, or could they also contract separately with large customers or across borders? Countries should therefore make it clear that the central purchasing function of the national utility does not imply exclusivity. IPPs should be permitted to seek their own customers.

Hybrid power markets will not disappear from the African landscape in the near future. To maximize their benefit, African governments and their development partners must establish a robust institutional foundation for the single-buyer model with clear criteria for off-take agreements. They must also improve their planning capabilities, establish clear policies for allocating new investment opportunities among the state-owned utilities and IPPs, and commit to competitive and timely bidding processes. Table 4.2 provides a list of common policy questions in the sector and corresponding solutions.

Development finance institutions and bilateral donors can provide advice and expertise to governments and utilities on establishing transparent frameworks and procedures for contracting and reaching financial closure with project sponsors and private investors. Yet they must be careful to pay sufficient attention to the peculiarities of the hybrid market. Otherwise lending to public utilities may unintentionally deepen hybrid markets' inherent contradictions and crowd out private investment. Above all, the sector requires stronger public institutions that can engage effectively with the private sector.

Table 4.2   Common Questions in Hybrid Power Markets and Their Policy Solutions

Question	Policy options
Who is responsible for security and adequacy of supply?	Develop standard for security and adequacy of supply. (The U.S. standard is one cumulative day of outage per 10 years; one day per year may be reasonable for countries in Sub-Saharan Africa.) Assign responsibility for reporting to utilities and monitoring supply adequacy to regulator.
Who is responsible for generation expansion planning?	Assign responsibility to ministry, regulator, or utility. Superior access to resources and professional staff may make the national utility the pragmatic choice, but this will require governance mechanisms to provide oversight and guidance on planning assumptions and criteria. Planning should be indicative, dynamic, flexible, and regularly updated, not a rigid master plan.
How are investment opportunities in new generation allocated between the national utility and IPPs?	Establish clear and transparent criteria for allocating new investment opportunities to either national utility or IPPs (for example, according to fuel source, technological expertise, or financing or contracting capability).
Who is responsible for initiating procurement of new generation plant and when?	Establish a procurement function (either in a PPP unit or linked to system operator or transmission function) that is informed by needs identified in planning process. Ensure adequate governance and oversight to ensure timely initiation of fair and transparent procurement.
Is competitive bidding required, or can unsolicited offers be considered and, if so, how?	Employ international competitive bidding processes whenever possible. Establish under what circumstances and how unsolicited bids can be considered.
Who is responsible for contracting IPPs?	Clarify market structure. Establish nonexclusive central purchasing function (possibly attached to system operator or transmission) that aggregates demand and signs PPAs. Build local capacity to negotiate effectively with private investors. Allow willing buyer-seller contracts between IPP and large customers and cross-border trades and contracting.
Can IPP PPA costs be passed on by national utility to customers?	Establish clear cost recovery mechanism for national utilities with captive customers who contract with IPPs and decide when PPA costs can be passed on to customers. Test competitiveness of procurement.
Will IPPs be fairly dispatched by the incumbent state-owned utility?	Ensure that PPAs, grid codes, and market rules have fair take or pay and dispatch provisions.

Source: Authors.
Note: IPP = independent power project; PPA = power-purchase agreement.

Hybrid power markets, with the incumbent state-owned utility designated as the single buyer of electricity from IPPs, have become the most common industry structure in Africa. Although the national utility can play a useful role in aggregating demand and entering into long-term contracts with new investors, few advantages are found in assigning it *exclusive* buying rights. Instead, IPPs should be able to enter into willing seller-buyer arrangements and supply directly to both the national utility and large customers. Large customers should also have choice and should be able to contract directly with IPPs or import power. Such an arrangement would require nondiscriminatory access to the grid. Perhaps a better description of such a model is a *central nonexclusive* buyer rather than a *single* buyer.

Thought also needs to be given to the long-term implications of signing 25- or 30-year contracts with IPPs. It may be advantageous to migrate to a more short-term market in the future. Including sunset clauses in PPAs would encourage IPPs to trade at least part of their production on a power exchange in the future.

The Possible Need to Redesign Regulatory Institutions

Most countries in Sub-Saharan Africa have established nominally independent regulatory agencies for their power sector. Regulation was originally intended to ensure financial viability, attract new investment, and encourage efficient, low-cost, and reliable service provision. Governments hoped that independent regulation would insulate tariff setting from political influence and improve the climate for private investment through more transparent and predictable decision making.

An analysis of data collected in the initial sample of 24 AICD countries indicates that the power sector performs better in countries with regulators than those without (figure 4.3). Yet the same countries show no obvious improvements in cost recovery, T&D losses, or reserve margins. These apparent contradictions can be explained. Cost recovery calculations can vary based on numerous assumptions that may affect estimates, and reporting on T&D losses is not always reliable. Furthermore, countries that lack regulators (such as Benin, Burkina Faso, Chad, the Democratic Republic of Congo, Mozambique, and Sudan) are among the poorest on the continent and face many additional challenges that affect the performance of their power sectors.

Despite the better performance of countries with regulators, it is far from clear whether regulation has catalyzed new private investment.

Figure 4.3   Power Sector Performance in Countries with and without Regulation

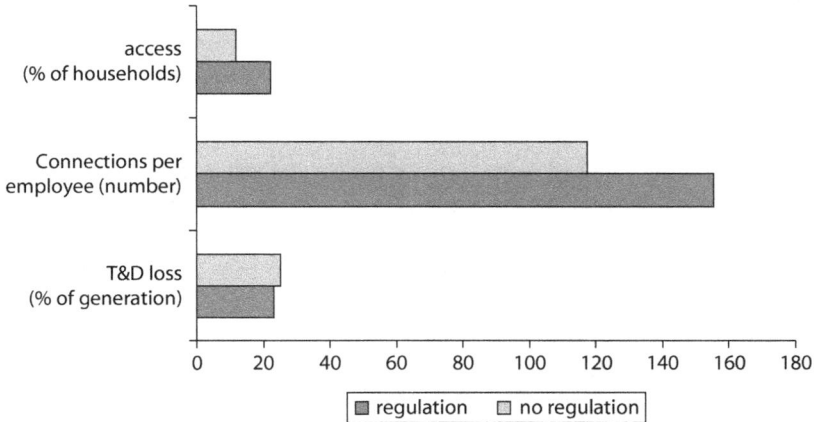

Source: Vagliasindi and Nellis 2010.
Note: T&D = transmission and distribution.

Some critics argue that regulatory agencies have exacerbated the very problems that they were meant to address while creating regulatory risk for investors. Inexperienced regulators tend to make unpredictable or noncredible decisions. Alternatively regulators may have been given excessively wide discretion and overly broad objectives and must make difficult decisions with important social and political consequences (Eberhard 2007).

The Challenges of Independent Regulation
Utility regulation in developing countries has clearly coincided with the emergence of new problems. In many cases, regulators are far from independent and are subject to pressure from governments to modify or overturn decisions. Turnover among commissioners has been high, with many resigning under pressure before completing their full term. The disconnect between law (or rule) and practice is often wide. Tariff setting remains highly politicized, and governments are sensitive to popular resentment against price increases, which are often necessary to cover costs. Establishing independent regulatory agencies may be particularly risky for all stakeholders (governments, utilities, investors, and customers) in sectors that are being reformed, especially when prices are not already high enough to ensure sufficient revenue. In some ways, it is not surprising to find political interference and pressure on regulators.

Governments in developing countries often underestimate the difficulty of establishing new public institutions. Building enduring systems of governance, management, and organization and creating new professional capacity are lengthy processes. Many regulatory institutions in developing countries are no more than a few years old, and few are older than 10. Many are still quite fragile and lack capacity.

Independent regulation requires strong regulatory commitment and competent institutions and people. The reality is that developing countries are often only weakly committed to independent regulation and face capacity constraints (Trémolet and Shah 2005). It may be prudent in such cases to acknowledge that weak regulatory commitment, political expediency, fragile institutions, and capacity constraints necessitate limits on regulatory discretion. This does not imply that independent regulation is undesirable. Because of limited institutional capacity in the sector, however, complementary, transitional, or hybrid regulatory options and models (such as regulatory contracts or outsourcing of regulatory functions) may be a better starting point.

Regulation by Contract

Most of the Sub-Saharan countries that were previously British colonies have independent regulators that operate within a system of common law with wide discretionary powers over decision making. On the other hand, those countries that were previously French colonies have tended to rely on regulatory contracts. For example, Cameroon, Côte d'Ivoire, Gabon, and Mali all have electricity concession contracts that incorporate core regulatory functions.

Regulatory contracts comprise detailed predetermined regimes (including multiyear, tariff-setting systems) in legal instruments such as basic law, secondary legislation, licenses, concession contracts, and PPAs (Bakovic, Tenenbaum, and Woolf 2003). They are generally constructed for private participation but may also be used to improve the performance of state-owned utilities.

Long-term contracts must accommodate for the possibility of unexpected events. In the French legal tradition, a general legal framework and an understanding between the parties to facilitate renegotiation is used to restore financial sustainability in extraordinary circumstances. On the other hand, the English legal tradition usually dictates specifying in advance the events that will trigger renegotiation.

Regulatory agencies can successfully coexist with incomplete regulatory contracts that require additional regulatory mechanisms. The law or contract could explicitly define the role of the regulator—for example, in periodic tariff setting, monitoring of performance, or mediation and arbitration. The regulator can also enhance the transparency of regulatory contracts by collecting, analyzing, and publishing performance data. Uganda provides a good example of successful coexistence of the two regulatory forms. The country has an independent regulator, but the generation and distribution components of the power sector have been privatized in concession agreements. Nevertheless, merging these two distinct legal traditions can create problems. For example, even if a contract specifies a tariff-setting formula, the regulator might feel obligated by its legislative mandate to intervene in the public interest. In these cases, clarifying regulatory roles and functions is essential.

Outsourcing Regulatory Functions

Countries may also outsource regulatory functions to external contractors, who perform tariff reviews, benchmarking, compliance monitoring, and dispute resolution. Power sectors that are beset by challenges or problems relating to a regulator's independence, capacity, or legitimacy are good candidates for regulatory outsourcing. The same is true for regulatory contracts that need additional support for effective administration. For example, the electricity concession in Gabon relies on external parties to monitor and verify performance indicators specified in its contract. Outsourcing might also be used when it is cost effective (Trémolet, Shukla, and Venton 2004).

Two main models of regulatory outsourcing are found. The first involves hiring outside consultants to provide technical support to regulators or the parties subject to a regulatory contract. Governments can also contract separate advisory regulators or expert panels, funded from an earmarked budget outside the line ministry. The strongest version of the second model requires the advisory regulator or expert panel to clearly explain its recommendations in publicly available documents. The sector minister (or other relevant authority) may request reconsideration of the recommendations but must do so within a specified period. If the minister rejects or modifies the recommendations, he or she must provide a written public explanation. Otherwise, the recommendations are enacted. Any policy directives or other communications from the minister to the regulator or expert panel must be made publicly available. The regulator or expert panel holds public consultations with any stakeholders affected by its recommendations (Brown and others 2006).

Governments may also hire expert panels to arbitrate disputes between regulators and utility operators or those arising from contested interpretations in regulatory contracts. Unlike conventional arbitration mechanisms, expert panels have the specialist expertise needed to analyze comprehensive tariff reviews and use procedures that are less formal and adversarial.

Regional economic bodies or regulatory associations could use expert panels to provide technical assistance to numerous national regulators. They would also provide greater continuity and consistency in specialist support and assist in harmonizing regulatory regimes, which would aid the integration of regional networks.

Toward Better Regulatory Systems

The different regulatory models embody varying degrees of regulatory discretion, but they are not mutually exclusive and often coexist (figure 4.4). How can countries choose among these options or decide on the appropriate combination?

Some observers have argued that the fundamental challenge in regulatory design is to find governance mechanisms that restrain regulatory discretion over substantive issues such as tariff setting (Levy and Spiller 1994). Others argue that some regulatory discretion is inevitable, or even desirable. The challenge is therefore to establish governance arrangements and procedures that allow a "nontrivial degree of bounded and accountable discretion" (Stern and Cubbin 2005).

A Model to Fit the Context

The context of a country's particular power sector should determine the level of regulatory discretion. Regulatory models and governance systems should be securely located within the political, constitutional, and legal arrangements of the country. They should also fit the country's levels of regulatory commitment, institutional development, and human resource capacity.

For a country with weak regulatory commitment and capacity, a good first step might be a set of low-discretion regulatory contracts without a regulatory agency (figure 4.5). In other countries with strong regulatory commitment but weak institutional development and capacity, regulatory functions could be contracted to an expert panel.

Countries with unique needs can also adopt a hybrid regulatory model. For example, a government could supplement an independent regulatory agency or regulatory contract by outsourcing some regulatory functions.

Figure 4.4   Coexistence of Various Regulatory Options

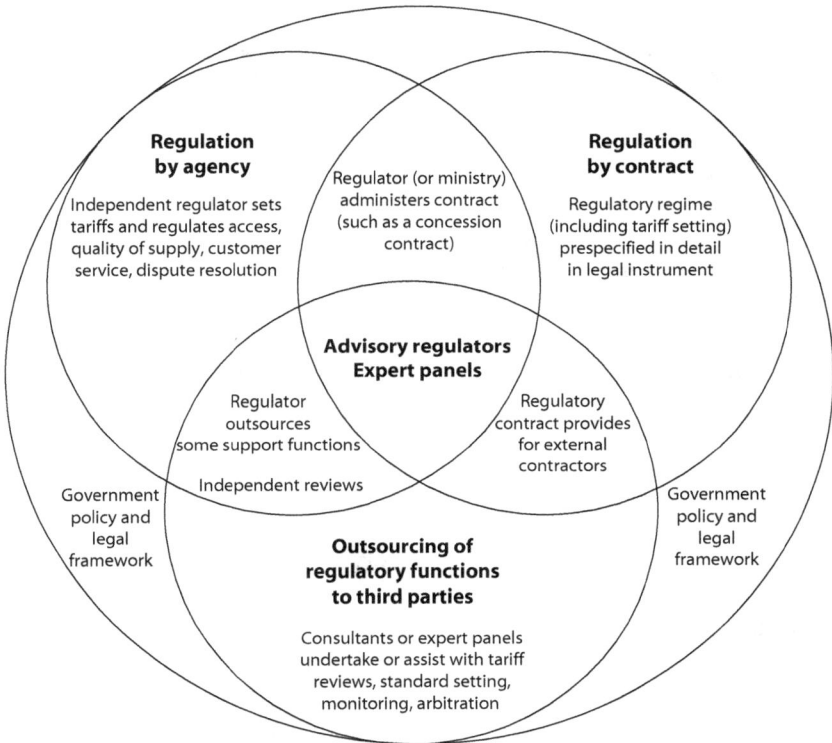

Regulation by agency

Independent regulator sets tariffs and regulates access, quality of supply, customer service, dispute resolution

Regulator (or ministry) administers contract (such as a concession contract)

Regulation by contract

Regulatory regime (including tariff setting) prespecified in detail in legal instrument

Advisory regulators Expert panels

Regulator outsources some support functions

Regulatory contract provides for external contractors

Independent reviews

Government policy and legal framework

Government policy and legal framework

Outsourcing of regulatory functions to third parties

Consultants or expert panels undertake or assist with tariff reviews, standard setting, monitoring, arbitration

Source: Eberhard 2007.

As noted, regulatory contracts can coexist with independent regulatory oversight.

Yet another possibility is a transitional path (as indicated in figure 4.5) in which the regulatory model adapts to accommodate changing circumstances. While regulatory commitment in a country grows, the government could contract strong advisory panels or establish a separate regulatory agency, perhaps with limited discretion at first. The responsibilities and functions of the regulatory agency could expand as sufficient institutional and resource capacity accumulates. Eventually, the government could outsource some regulatory functions.

No regulatory model is ideal, and a country's regulatory reform process may not always lead to a full-fledged independent regulatory agency. In fact, the context simply may not call for an independent regulator, and an

Figure 4.5   Choice of Regulatory Model Based on the Country Context

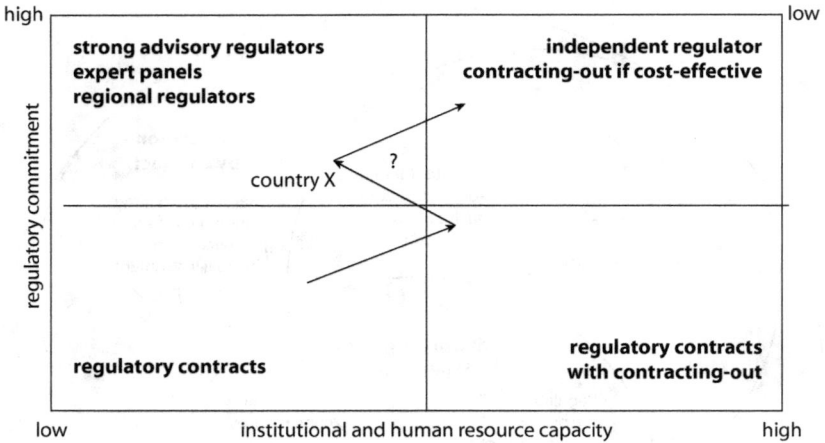

Source: Adapted from Brown and others 2006.

expert panel or a well-designed regulatory contract would suit the country's needs. Each country therefore must choose from a menu of regulatory options to create a hybrid model that best fit its particular situation. The model must be flexible enough to evolve according to growth in a country's regulatory commitment and capacity. In the end, designing and implementing legitimate, competent regulatory institutions in developing countries will always be a challenge. Nevertheless, establishing an effective regulatory system is essential to the region's strategy of increasing private participation in the power sector.

More effective regulation of incumbent state-owned utilities will remain a critical challenge. Regulators can play a useful role in ensuring that tariffs are cost reflective while improving efficiencies and encouraging utilities to reduce costs. Improved financial performance also helps utilities to raise private debt and fund capacity expansion. These issues are discussed further in chapters 6 and 7.

Notes

1. The only exception is a short-term energy market in the Southern African Power Pool. The quantities traded, however, are extremely small.
2. Uganda is one of the exceptions where generation, transmission, and distribution were fully unbundled. In Kenya, generation (KenGen) has been separated

from transmission and distribution (KPLC). Ghana has unbundled its transmission company and has a separate distribution company. Nigeria has technically unbundled its utility, although the separate entities still coordinate with each other. For historical reasons, local governments in Namibia and South Africa assume some responsibility for distribution.

3. The author of this section is John Nellis (2008).

4. Typically expressed as a loss-of-load probability and an associated generation-reserve margin.

Bibliography

AICD (Africa Infrastructure Country Diagnostic). 2008. *Power Sector Database.* Washington, DC: World Bank.

Bakovic, Tonci, Bernard Tenenbaum, and Fiona Woolf. 2003. "Regulation by Contract: A New Way to Privatize Electricity Distribution?" Energy and Mining Sector Board Discussion Paper Series, Paper 7, World Bank, Washington, DC.

Besant-Jones, J. E. 2006. "Reforming Power Markets in Developing Countries: What Have We Learned?" Energy and Mining Sector Board Discussion Paper Series, Paper 19, World Bank, Washington, DC.

Brown, Ashley C., Jon Stern, Bernard W. Tenenbaum, and Defne Gencer. 2006. *Handbook for Evaluating Infrastructure Regulatory Systems.* Washington, DC: World Bank.

Clark, Alix, Mark Davis, Anton Eberhard, and Njeri Wamakonya. 2005. "Power Sector Reform in Africa: Assessing the Impact on Poor People." ESMAP Report 306/05, World Bank, Washington, DC.

Eberhard, A. 2007. "Matching Regulatory Design to Country Circumstances: The Potential for Hybrid and Transitional Models." Gridlines Note 23, PPIAF, World Bank, Washington, DC.

Gboney, William K. 2009. "Econometric Assessment of the Impact of Power Sector Reforms in Africa: A Study of the Generation, Transmission and Distribution Sectors." Thesis to be submitted in fulfillment of the PhD degree, City University, London.

Ghanadan, R., and A. Eberhard. 2007. "Electricity Utility Management Contracts in Africa: Lessons and Experience from the TANESCO-NET Group Solutions Management Contract in Tanzania." MIR Working Paper, Management Program in Infrastructure Reform and Regulation, Graduate School of Business, University of Cape Town, Cape Town, South Africa.

Gratwick, K. N., and Anton Eberhard. 2008. "An Analysis of Independent Power Projects in Africa: Understanding Development and Investment Outcomes." *Development Policy Review* 26 (3): 309–38.

Levy, B., and P. Spiller. 1994. "The Institutional Foundations of Regulatory Commitment: A Comparative Analysis of Telecommunications Regulation." *Journal of Law, Economics and Organization* 10 (1): 201–46.

Nellis, John. 2008. "Private Management Contracts in Power Sector in Sub-Saharan Africa." Internal note, Africa Infrastructure Country Diagnostic, World Bank, Washington, DC.

Stern, Jon, and John Cubbin. 2005. "Regulatory Effectiveness: The Impact of Regulation and Regulatory Governance Arrangements on Electricity Industry Outcomes." Policy Research Working Paper Series 3536, World Bank, Washington, DC.

Trémolet, Sophie, and Niraj Shah. 2005. "Wanted! Good Regulators for Good Regulation." Unpublished research paper, PPIAF, World Bank, Washington, DC.

Trémolet, Sophie, Padmesh Shukla, and Courtenay Venton. 2004. "Contracting Out Utility Regulatory Functions." Unpublished research paper, PPIAF, World Bank, Washington, DC.

Vagliasindi, Maria, and John Nellis. 2010. "Evaluating Africa's Experience with Institutional Reform for the Infrastructure Sectors." AICD Working Paper 22, World Bank, Washington, DC.

World Bank. 2007. *Private Participation in Infrastructure (PPI) Database.* Washington, DC: World Bank.

CHAPTER 5

Widening Connectivity and Reducing Inequality

Coverage of electricity services in Sub-Saharan Africa, stagnant over the past decade, skews strongly toward higher-income households and urban areas. Many of those who remain without a connection live reasonably close to existing networks, which suggests that in addition to supply constraints, demand-side barriers may be a factor. In these circumstances, the key questions are whether African households can afford to pay for modern infrastructure services such as electricity—and, if not, whether African governments can afford to subsidize them.

The business-as-usual approach to expanding service coverage in Africa does not seem to be working. Reversing this situation will require rethinking the approach to service expansion in four ways. First, coverage expansion is not just about network rollout. A need exists to address demand-side barriers such as high connection charges. Second, it is important to remove unnecessary subsidies to improve cost recovery for household services and ensure that utilities have the financial basis to invest in service expansion. Third, it is desirable to rethink the design of utility subsidies to target them better and to accelerate service expansion. Fourth, progress in rural electrification cannot rely only on decentralized options; it requires a sustained effort by national utilities supported by systematic planning and dedicated rural electrification funds (REFs).

Low Electricity Connection Rates

Coverage of electricity services in Africa is very low by global standards. Connection rates are less than 30 percent in Africa, compared with approximately 65 percent in South Asia and more than 85 percent in East Asia and the Middle East. Africa's low coverage of infrastructure services to some extent reflects its relatively low urbanization rates, because urban agglomeration greatly facilitates the extension of infrastructure networks.

Household surveys show only modest gains in access to modern infrastructure services over 1990–2005 (figure 5.1). The overall trend masks the fact that the percentage of households with connections in urban

Figure 5.1   Patterns of Electricity Service Coverage in Sub-Saharan Africa

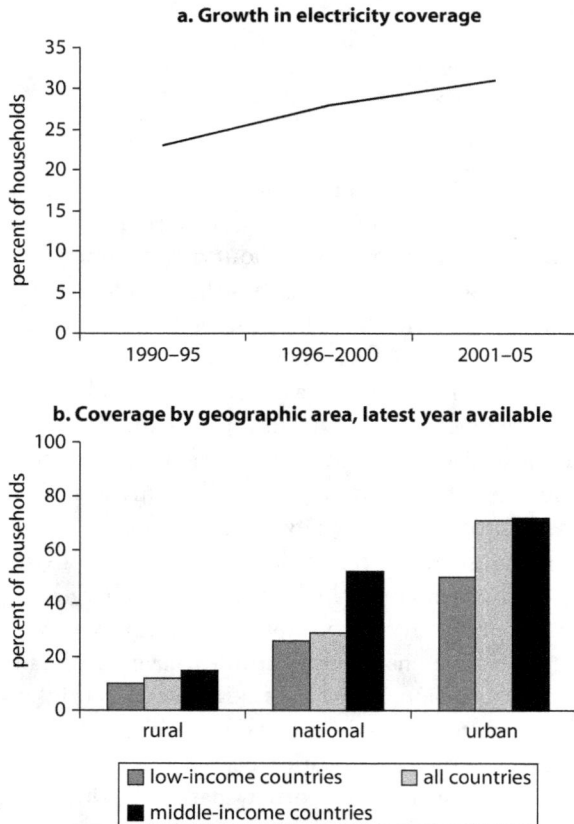

a. Growth in electricity coverage

b. Coverage by geographic area, latest year available

Source: Banerjee and others 2008; Eberhard and others 2008.

areas has actually declined. Although many new connections are being made in urban areas, declining urban coverage largely reflects service providers' inability to keep pace with average urban population growth of 3.6 percent a year.

The pace of service expansion differs across countries. The most dramatic increase in electricity connections was seen in South Africa after the advent of democracy in 1994. Coverage increased from approximately one-third of the population to more than two-thirds in less than a decade (Marquard and others 2008). A few countries—such as Cameroon, Côte d'Ivoire, Ghana, and Senegal—have made some progress, and close to half of their people now have access. (Box 5.1 examines Ghana's electrification program.) These are exceptions, however, and most countries of Sub-Saharan Africa lag far behind. For example, Uganda's electrification rate stands at 8 percent and Chad's at 4 percent (figure 5.2).

Mixed Progress, despite Many Agencies and Funds

Despite accelerating urbanization, the region's rural areas still account for approximately two-thirds of the total population, which presents significant challenges in raising access rates. It is obviously cheaper to electrify urban areas, followed by higher-density rural areas. Off-grid technologies such as solar photovoltaic panels become an option in remote areas but are still very expensive—typically $0.50–0.75 per kilowatt-hour (kWh). Minigrids, where feasible, are more attractive options in remote areas, especially when combined with small-scale hydropower facilities (ESMAP 2007).

Incumbent national utilities—mostly state owned and vertically integrated—are responsible for urban (and often rural) electrification. A significant trend during the past decade, however, has been the establishment of special-purpose agencies and funds for rural electrification. Half the countries in the Africa Infrastructure Country Diagnostic (AICD) sample have REAs (rural electrification agencies), and more than two-thirds have dedicated REFs. Funding sources for REFs may be levies, fiscal transfers, donor contributions, or combinations of these. The majority of countries have full or partial capital subsidies for rural connections and explicit planning criteria (usually population density, least cost, or financial or economic returns). In some cases, political pressures trump these criteria.

How effective have these institutional and funding mechanisms been in accelerating rural electrification? On average, greater progress has been

Box 5.1

Ghana's Electrification Program

Ghana boasts a national electrification rate of nearly 50 percent. Urban rates of access hover near 80 percent, and rural rates at approximately 20 percent. With access of the population to electricity at less than 25 percent in the region, Ghana's recent electrification experience may be instructive for neighboring countries.

Starting in 1989, when Ghana's access rates were estimated at 20 percent and the grid supply covered only one-third of the country's land area, electrification efforts were intensified under the National Electrification Scheme (NES), which was designed to connect all communities with a population of more than 500 to the national grid between 1990 and 2020.

The National Electrification Master Plan subsequently laid out 69 projects that would span 30 years to realize the stated policy goal. The first two five-year phases of the plan were undertaken between 1991 and 2000; the country's two state-owned utilities, Electricity Company of Ghana and the Volta River Authority, were charged with implementation. A rural electrification agency was not used. Project costs of $185 million were covered largely via concessionary financing from several multilateral and bilateral donors.

In addition to the central role of the utilities and the prominence of concessionary lending, the Self-Help Electrification Programme (SHEP) was noteworthy in advancing the aims of the NES. SHEP was the means by which communities, within a certain proximity to the network and otherwise not targeted for near-term electrification, were able to be connected by purchasing low-voltage distribution poles and demonstrate the readiness of a minimum number of households and businesses to receive power. SHEP was further supported by a 1 percent levy on electricity tariffs.

As of 2004, efforts under the NES had led to the electrification of more than 3,000 communities. Contrary to expectations, however, an indigenous industry to supply products for the electrification program has not taken off. Furthermore, SHEP is now considered defunct, having been unable to sustain itself financially. Nevertheless, the NES continues and is cofinanced by development finance institutions and local Ghanaian banks, with an increasing emphasis on minigrids and standalone systems.

Source: Clark and others 2005; Mostert 2008.

Figure 5.2   Electrification Rates in the Countries of Sub-Saharan Africa, Latest Year Available

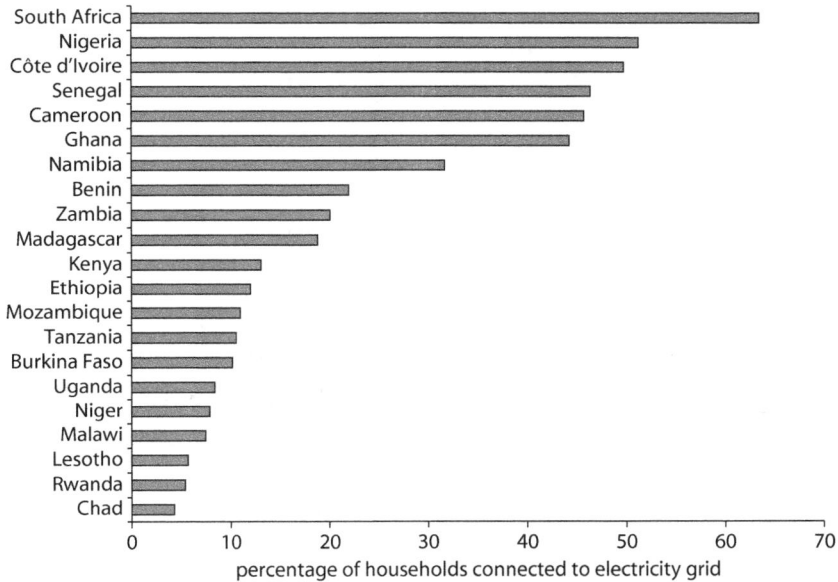

percentage of households connected to electricity grid

Source: Eberhard and others 2008.

made in those countries with electrification agencies and especially those with dedicated funds (figure 5.3). Having a clear set of electrification criteria also makes a difference.

Countries with higher urban populations also tend to have higher levels of rural electrification, because urban customers tend to cross-subsidize rural electrification (figure 5.4). Surprisingly, no correlation could be found between the proportion of utility income derived from nonresidential electricity sales and the level of growth in residential connections. One would have expected that increased revenue from industrial and commercial customers would also allow for the cross-subsidization of rural electrification.

A recent review of electrification agencies in Africa has concluded that centralized approaches, in which a single utility is responsible for national rural electrification, for the most part have been more effective than decentralized approaches involving several utilities or private companies, provided the national utility is reasonably efficient (Mostert 2008).

Figure 5.3 Rural Electrification Agencies, Funds, and Rates in Sub-Saharan Africa

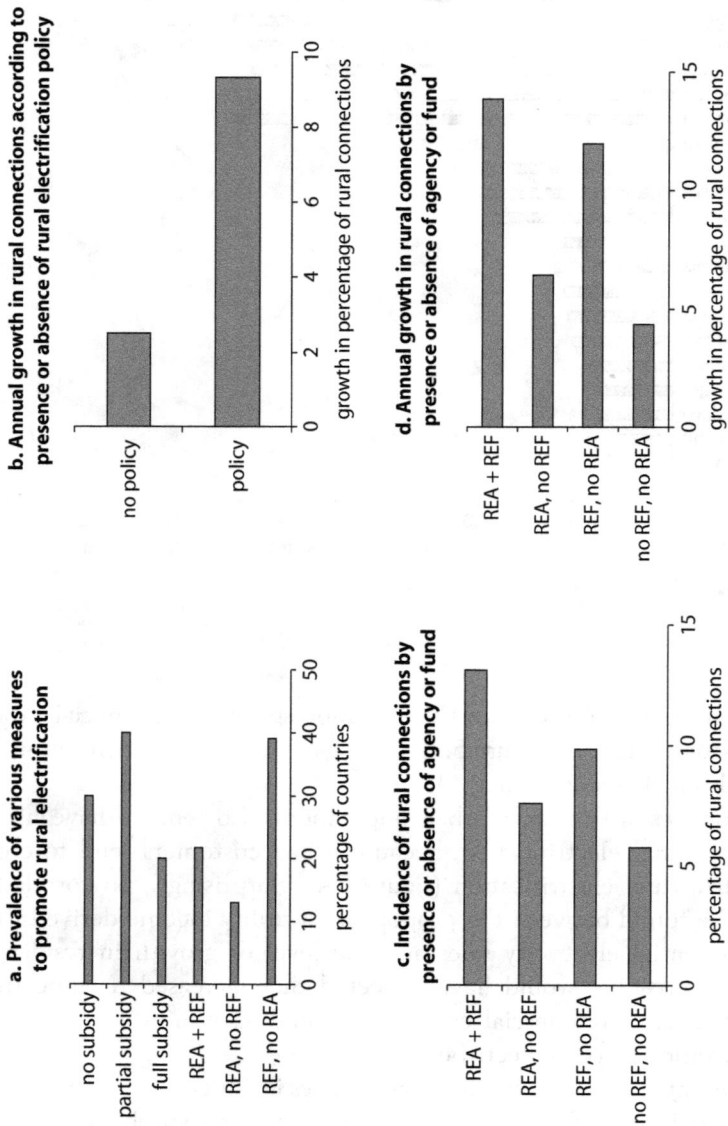

a. Prevalence of various measures
to promote rural electrification

b. Annual growth in rural connections according to
presence or absence of rural electrification policy

c. Incidence of rural connections by
presence or absence of agency or fund

d. Annual growth in rural connections by
presence or absence of agency or fund

Source: Eberhard and others 2008.
Note: REA = rural electrification agency; REF = rural electrification fund. Annual growth in new connections may seem high but comes off a low base; the overall percentage increase in households with access remains low.

Figure 5.4   Countries' Rural Electrification Rates by Percentage of Urban Population

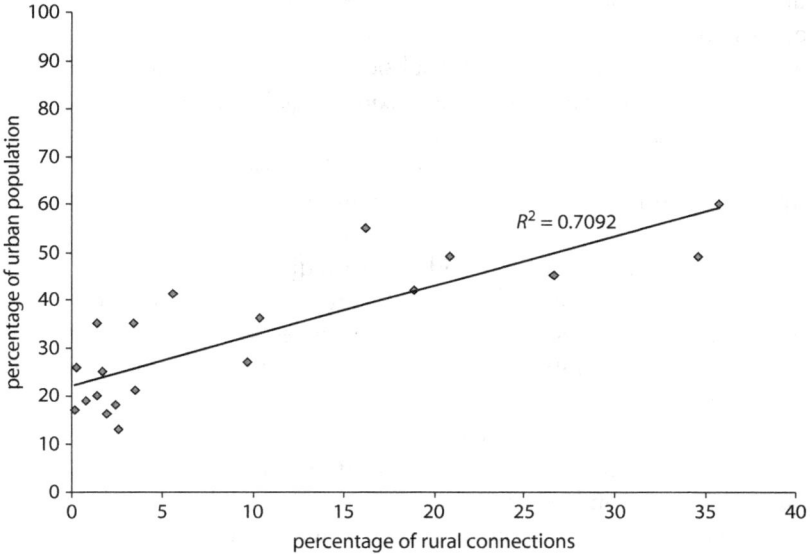

Source: Eberhard and others 2008.

Côte d'Ivoire and Ghana are examples of countries that have made good progress with a centralized approach to rural electrification. South Africa has also relied mainly on its national utility, Eskom, to undertake rural electrification, with considerable success. In contrast, countries such as Burkina Faso and Uganda have made slow progress, and rural electrification rates remain very low. These are obviously very poor countries, but it is also noteworthy that they have allowed their REFs to recruit multiple private companies on a project-by-project basis rather than make their national utilities responsible for extending access. Exceptions may be identified, however; for example, decentralized rural electrification has been more successful in Mali and Senegal.

At first glance, the findings of the Mostert study (2008) would appear to contradict our previous findings that countries with electrification funds (and, to a lesser extent, agencies) tend to perform better in electrification. It should be noted, however, that Mostert's categorization of countries that rely on central utilities for electrification, on the one hand, versus those with REFs and REAs, on the other, does not match the situation in many countries where the two approaches complement one another. For example, South Africa has an electrification fund, but Eskom is responsible for rural electrification. The purpose of the fund is to ring-fence subsidy sources from commercial revenue earned by the utility. Electrification funds create transparency around subsidies and thus help avoid situations where utilities face mixed social and commercial incentives.

Decentralized rural electrification often makes most sense when applied to the implementation of off-grid projects and as a way of exploiting the private initiatives of small-scale entrepreneurs and motivated communities. Mostert (2008) cites successful examples of this approach in Ethiopia, Guinea, and Mozambique. The lesson is that it may be unrealistic to allocate responsibility for all electrification to separate electrification agencies, but that these agencies should focus mainly on minigrid or off-grid options that complement the efforts of the main utility charged with extending grid access.

Universal access to electricity services is still many decades away for most countries in Sub-Saharan Africa. By projecting current service expansion rates forward and taking into account anticipated demographic growth, it is possible to estimate the year during which countries would reach universal access to each of the modern infrastructure services. The results are sobering. Under business as usual, fewer than 45 percent will reach universal access to electricity in 50 years (Banerjee and others 2008).

Inequitable Access to Electricity

Electricity coverage in Sub-Saharan Africa is low and skewed to more affluent households. Coverage varies dramatically across households with different budget levels (figure 5.5). Among the poorest 40 percent of the population, coverage of electricity services is well below 10 percent. Conversely, the vast majority of households with coverage belong to the more affluent 40 percent of the population. In most countries, inequality of access has increased over time, which suggests that most new connections have gone to more affluent households. This is not entirely surprising, given that even among households with greater purchasing power, coverage is far from universal.

The coverage gap for urban electricity supply is about demand as much as supply. For electricity, the power infrastructure is physically close to 93 percent of the urban population, but only 75 percent of those connect to the service (table 5.1). As a result, approximately half the population without access to the service lives close to power infrastructure, and the

Figure 5.5   For the Poorest 40 Percent of Households, Coverage of Modern Infrastructure Services Is below 10 Percent

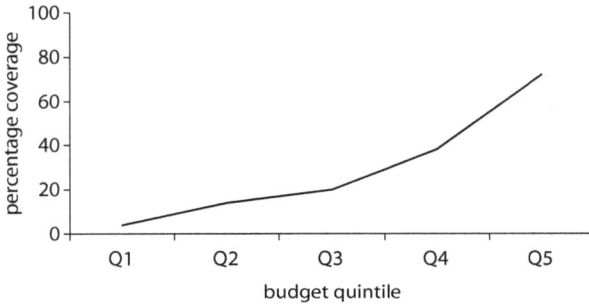

Source: Banerjee and others 2008.

Table 5.1   Proportion of Infrastructure Electricity Coverage Gap in Urban Africa Attributable to Demand and Supply Factors

| | Percentage, population-weighted average | | | | |
| | Decomposition of coverage | | | Proportion of gap attributed to: | |
	Access	Connection	Coverage	Supply	Demand
Low-income countries	93	73	69	50	50
Middle-income countries	95	86	81	39	61
Overall	93	75	71	48	52

Source: Banerjee and others 2008.
Note: Access is defined as the percentage of the population that lives physically close to infrastructure. Connection is defined as the percentage of the population that connects to infrastructure when it is available. Coverage is defined as the percentage of the population that has the infrastructure service; it is essentially the product of access and connection. In calculating the distribution of the infrastructure coverage gap attributable to demand and supply factors, the connection rate of the top budget quintile in each geographical area is taken to be an upper bound on potential connection absent demand-side constraints.

coverage gap is as much about demand (affordability) as supply. This phenomenon can often be directly observed in African cities where informal settlements flanking major road corridors lack power service even though distribution lines are running overhead.

It may appear paradoxical that households do not universally take up connections to modern infrastructure services once networks become physically available, but often clear budget constraints are present. Poor households cannot afford high connection charges and rely instead on more accessible substitutes such as wood fuel, charcoal, kerosene, and bottled gas. Of course, slow progress in connections to electricity distribution networks cannot be explained only by demand or affordability constraints: Poorly performing utilities also have large backlogs in connecting users who are willing to pay.

The tenure status of households may also impede connection to modern infrastructure services. A study of slum households in Dakar and Nairobi finds that electricity coverage is more than twice as high among owner occupiers as among tenants. Even among owner occupiers, lack of formal legal titles can also affect connection to services (Gulyani, Talukdar, and Jack 2008).

Affordability of Electricity—Subsidizing the Well-Off

African households get by on very limited household budgets. The average African household of five persons has a monthly budget of less than $180; the range is from nearly $60 in the poorest quintile to $340 in the richest quintile (table 5.2). Thus, even in Africa's most affluent households, purchasing power is fairly modest in absolute terms. Across the spectrum, household budgets in middle-income countries are roughly twice those in low-income countries.

Expenditure on infrastructure services absorbs a significant share of the nonfood budget. Most African households spend more than half of their modest budgets on food, with little left over for other items. Spending on infrastructure services (including utilities, energy, and transport) averages 7 percent of a household's budget, though in some countries this can be 15–25 percent. As household budgets increase, infrastructure services absorb a growing share and rise from less than 4 percent among the

Table 5.2   Monthly Household Budget
2002 dollars

		Income group				
	National	Poorest quintile	Second quintile	Third quintile	Fourth quintile	Richest quintile
Overall	177	59	97	128	169	340
Low-income countries	139	53	80	103	135	258
Middle-income countries	300	79	155	181	282	609

Source: Banerjee and others 2008.

poorest to more than 8 percent among the richest (figure 5.6). In terms of absolute expenditure, this difference is even more pronounced: Whereas households in the poorest quintile spend on average no more than $2 per month on all infrastructure services, households in the richest quintile spend almost $40 per month.

Given such low household budgets, a key question is whether households can afford to pay for modern infrastructure services. One measure of affordability is nonpayment for infrastructure services. Nonpayment directly limits the ability of utilities and service providers to expand networks and improve services by undermining their financial strength. From household surveys, it is possible to compare for each quintile the percentage of households that report paying for the service with the percentage of households that report using the service. Those that do not pay include clandestine collections and formal customers who fail to pay their bills. Overall, an estimated 40 percent of people connected to infrastructure services do not pay for them. Nonpayment rates range from approximately 20 percent in the richest quintile to approximately 60 percent in the poorest quintile (figure 5.7). A significant nonpayment rate, even among the richest quintiles, suggests problems of payment culture alongside any affordability issues.

The cost of a monthly subsistence consumption of power can range from $2 (based on a low-cost country tariff of $0.08 per kWh and an

Figure 5.6   Infrastructure Services Absorb More of Household Budgets as Incomes Rise

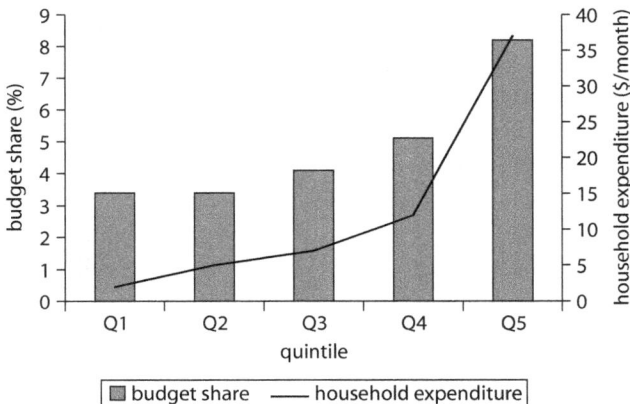

Source: Banerjee and others 2008.

Figure 5.7   About 40 Percent of Households Connected Do Not Pay

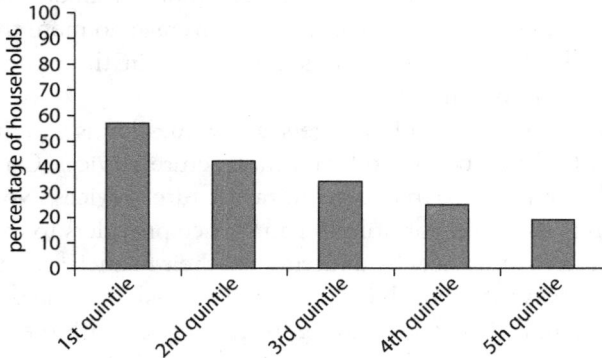

Source: Banerjee and others 2008.

absolute minimum consumption of 25 kWh) to $8 (based on a high-cost country tariff of $0.16 per kWh and a more typical modest household consumption of 50 kWh) (figure 5.8).

An affordability threshold of 3 percent of household budgets gauges what utility bills might be affordable to African households. By looking at the distribution of household budgets, it is possible to calculate the percentage of households for whom such bills would absorb more than 3 percent of their budgets and thus prove unaffordable. Monthly bills of $2 are affordable for almost the entire African population. Monthly bills of $8 would remain affordable for most of the population of the middle-income African countries.

In low-income countries, monthly bills of $8 would remain perfectly affordable for the richest 20–40 percent of the population, the only ones enjoying access. They would not be affordable, however, for the poorest 60–80 percent that currently lack access if services were extended to them. The affordability problems associated with a universal access policy would be particularly great for a handful of the poorest low-income countries—Burundi, the Democratic Republic of Congo, Ethiopia, Guinea-Bissau, Malawi, Niger, Tanzania, and Uganda—where as much as 80 percent of the population would be unable to afford a monthly bill of $8.

Detailed analysis of the effect of significant tariff increases of 40 percent for power in Mali and Senegal confirms that the immediate poverty impact on consumers is small, because very few poor consumers are connected to the service. However, broader poverty impacts may be seen as

Figure 5.8   Subsistence Consumption Priced at Cost Recovery Levels Ranges from $2 to $8

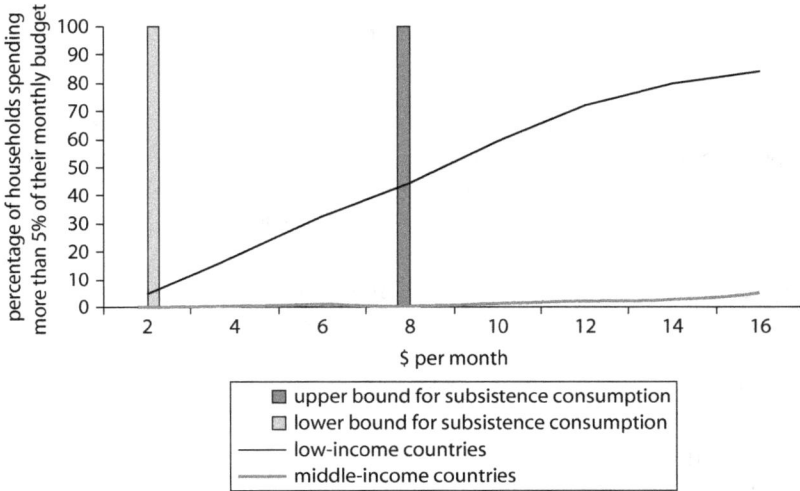

Source: Banerjee and others 2008.

the effects of higher power prices work their way through the economy, and these second-round effects on wages and prices of goods in the economy as a whole can be more substantial (Boccanfuso, Estache, and Savard 2009; Boccanfuso and Savard 2000, 2005).

Notwithstanding these findings, tariffs for power are heavily subsidized in most African countries. On average, power tariffs recover only 87 percent of full costs. The resulting implicit service subsidies amount to as much as $3.6 billion a year, or 0.56 percent of Africa's gross domestic product (GDP) (Foster and Briceño-Garmendia 2009).

Moreover, these subsidies largely bypass low-income households not even connected to services. Tariff structure design could help subsidize consumption by poor households (box 5.2). However, usually most of the resulting subsidy benefits the nonpoor. Because electricity subsidies are typically justified by the need to make services affordable to low-income households, a key question is whether subsidies reach such households.

Results across a number of African countries show that the share of subsidies going to the poor is less than half their share in the population, indicating a very pro-rich distribution (figure 5.9). This result is hardly surprising given that connections to power services are already highly skewed toward more affluent households. This targeting compares

Box 5.2

Residential Electricity Tariff Structures in Sub-Saharan Africa

Electricity tariff structures often take the form of increasing block tariffs (IBTs) in which a lower unit price is charged within the first consumption block and higher prices in subsequent consumption blocks. In contrast, decreasing block tariffs (DBTs) have lower unit charges for higher consumption-level blocks. Electricity tariff structures can also be linear, where the first unit of electricity consumed costs the same as the last unit consumed.

Block tariff schemes are commonly supplemented by fixed charges; the combination is known as two-part electricity tariffs. The fixed charge is usually determined by the level of development of the network, the location, service costs, and—when subsidization practice applies—the purchasing power of the consumer.

Two-thirds of the prevailing electricity tariff structures in Sub-Saharan Africa are IBTs, and one-third are single block or linear rates. The use of linear rates is more common in countries with prepayment systems such as Malawi, Mozambique, and South Africa.

About half the countries in Africa have adopted two-part tariffs that combine fixed charges with block energy pricing.

The conventional regulatory wisdom is that IBTs are designed as "lifeline" or "baseline" tariffs trying to align the first block of low consumption to a subsidized tariff and higher levels of consumption to higher pricing that would ultimately allow for cost recovery. This assumes that poorer customers will have lower consumption levels. This is a reasonable assumption in the power sector, where consumption is correlated with ownership of power-consuming devices, more of which are owned by wealthier households.

Two-thirds of African countries define the first block at 50 kWh/month or less. Countries in this group include Uganda, at 15 kWh/month; Cape Verde and Côte d'Ivoire, at 40 kWh/month; and Burkina Faso, Cameroon, Ethiopia, Kenya, and Tanzania, at 50 kWh/month. The Democratic Republic of Congo and Mozambique also define a modest threshold level for their first block (100 kWh). Ghana and Zambia have a large first block (300 kWh).

Source: Briceño-Garmendia and Shkaratan 2010.

Figure 5.9   Electricity Subsidies Do Not Reach the Poor

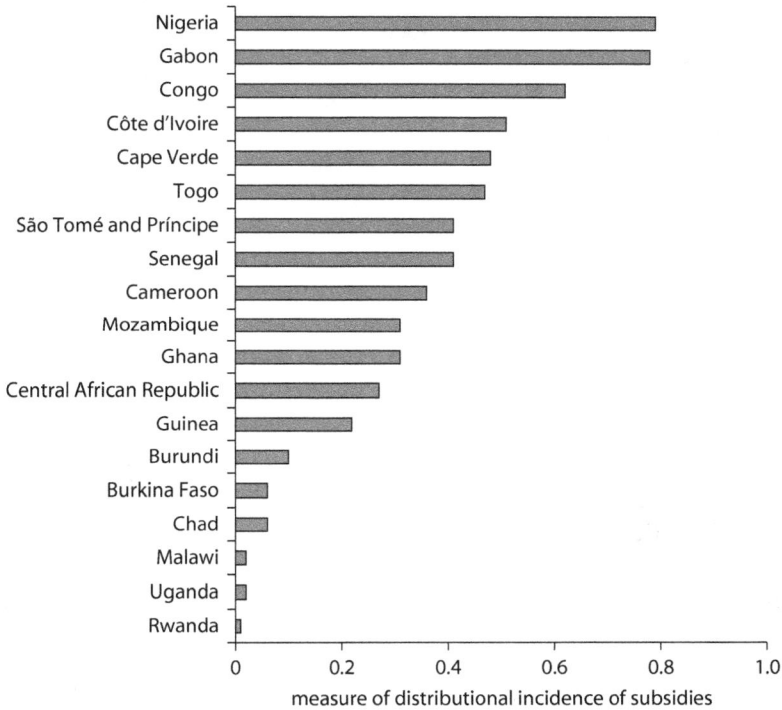

Source: Banerjee and others 2008; Wodon 2007a, b.
Note: A measure of distributional incidence captures the share of subsidies received by the poor divided by the proportion of the population in poverty. A value greater than one implies that the subsidy distribution is progressive (pro-poor), because the share of benefits allocated to the poor is larger than their share in the total population. A value less than one implies that the subsidy distribution is regressive (pro-rich).

unfavorably with other areas of social policy. To put these results in perspective, it is relevant to compare them with the targeting achieved by other forms of social policy. Estimates for Cameroon, Gabon, and Guinea indicate that expenditures on primary education and basic health care reach the poor better than power subsidies (Wodon 2007a).

Can African governments afford to further expand today's subsidy model to achieve universal access? There is little justification for utility subsidies at present given that they do not typically reach unconnected low-income households and that more affluent connected households do not need subsidies to afford the service. However, the preceding analysis indicated that affordability would become a major issue to the extent that

Africa's low-income countries move aggressively toward universal access. Given the very high macroeconomic cost today of subsidizing even the minority of the population with access to power, it is legitimate to question whether African governments can afford to scale up this subsidy-based model to the remainder of their populations.

Providing universal use of service, subsidies of $2 per household would absorb 1.1 percent of GDP over and above existing spending. This amount is high in relation to existing operations and maintenance expenditure, so it is difficult to believe that it would be affordable (figure 5.10).

The cost of providing a one-time capital subsidy of $200 to cover network connection costs for all unconnected households over 20 years would be substantially lower at 0.35 percent of GDP. A key difference is that the cost of this one-time subsidy would disappear at the end of the decade, whereas the use of a service subsidy would continue indefinitely.

The welfare case is quite strong for one-time capital subsidies to support universal connection. This is generally the most effective means of subsidizing the poor. Direct grants could also be made to indigent households, but effective targeting is difficult and administration complex. Cross-subsidies can also be achieved through the design of tariff structures that allow for lower rates for a "lifeline" amount of electricity usage for poor households. AICD data across a number of African countries

Figure 5.10   Subsidy Needed to Maintain Affordability of Electricity

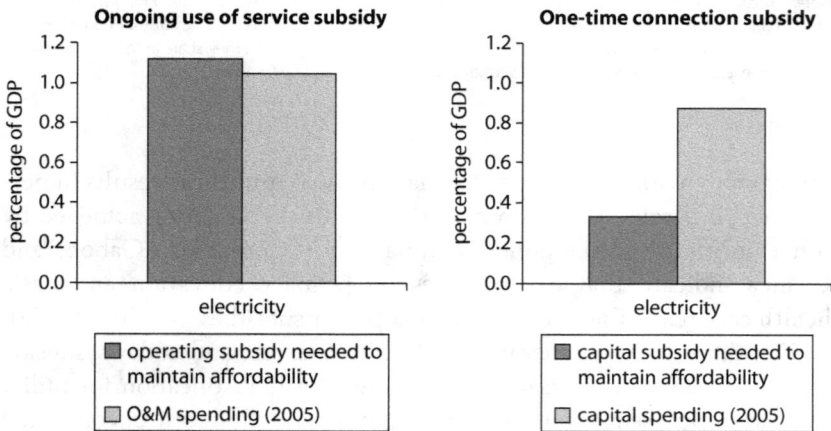

Source: Banerjee and others 2008.
Note: GDP = gross domestic product; O&M = operations and maintenance.

suggest that many current tariff structures are poorly designed. High fixed charges may inhibit affordability. The level and scope of the lifeline block in IBTs may also be inappropriate, giving too small a benefit to the poor. Alternatively, "pro-poor" tariffs may be poorly targeted and benefit wealthier consumers if the lifeline block is available too widely.

It is well known that households without access to utility services end up paying much higher prices, which limits their energy consumption to very low levels. The cost of providing basic illumination through candles is much more costly than electricity per effective unit of lighting.

Nonmonetary benefits of connection can also be very significant. Beyond the potential monetary savings, electricity coverage is associated with a wide range of health, education, and productivity benefits. For example, better electricity provision improves literacy and primary school completion rates, because better-quality light allows students to read and study in the absence of sunlight.

Policy Challenges for Accelerating Service Expansion

The business-as-usual approach to expanding service coverage in Africa does not seem to be working. The low and stagnant coverage of household services comes with a major social and economic toll. Under the business-as-usual approach, most African countries have tackled universal access by providing heavily subsidized services. This approach has tended to bankrupt and debilitate sector institutions without bringing about any significant acceleration of coverage. Furthermore, the associated public subsidies have largely bypassed most needy groups. Few services and countries are expanding coverage at rates high enough to outstrip demographic growth, particularly urbanization.

Reversing this situation will require rethinking the approach to service expansion in four ways. First, coverage expansion is not just about network rollout. There is a need to address demand-side barriers such as high connection charges or legal tenure. Second, it is important to remove unnecessary subsidies to improve cost recovery for household services and ensure that utilities have the financial basis to invest in service expansion. Third, it is desirable to rethink the design of utility subsidies to target them better and to accelerate service expansion. Fourth, progress in rural electrification cannot rely only on decentralized options; it requires a sustained effort by national utilities supported by systematic planning and dedicated REFs.

Don't Forget the Demand Side of the Equation

Overlooking the demand side of network rollout can lead to much lower returns on infrastructure investments. The challenge of reaching universal access is typically considered a supply problem of rolling out infrastructure networks to increasingly far-flung populations. Household survey evidence shows, however, that in urban areas, a significant segment of the unserved population lives close to a network.

The lower the connection rate to existing infrastructure networks, the lower the financial, economic, and social returns to the associated investment, because the physical asset is operating below its full carrying capacity. This finding has five implications for network rollout strategy.

First, connection, rather than access, needs to be considered the key measure of success. Projects that aim to expand service coverage too often measure their outcomes by the number of people who can connect to the network provided. As a result, little attention is given to whether these connections materialize after the project. Unless the focus of monitoring and evaluation shifts from access to connection, those involved in project implementation will have little incentive to think about the demand side of service coverage.

Second, the most cost-effective way of increasing coverage may be to pursue densification programs that aim to increase connection rates in targeted areas. Unserved populations living physically close to infrastructure networks could (in principle) be covered at a much lower capital cost than those living farther away, providing the highest potential return to a limited investment budget. In that sense, they may deserve priority attention in efforts to raise coverage.

Third, expanding coverage is not just about network engineering—it requires community engagement. Dealing with the demand-side barriers preventing connection requires a more detailed understanding of the utility's potential client base. What are their alternatives? How much can they afford to pay? What other constraints do they face? This, in turn, suggests a broader skill base than utilities may routinely engage, one that goes beyond standard expertise in network engineering to encompass sociological, economic, and legal analysis of—and engagement with—the target populations.

Fourth, careful thought should be given to how connection costs might be recovered. As noted previously, Africa's widespread high connection charges are one obvious demand-side barrier to connection, even when use-of-service charges would be affordable. In these circumstances, it is legitimate to ask whether substantial, one-time, upfront connection charges are the most sensible way to recover the costs of making network connections. Alternatives can be considered, including repaying connection costs over several years through an installment plan, socializing connection costs by recovering them through the general tariff and hence sharing them across the entire customer base, or directly subsidizing them from the government budget.

Fifth, expansion of utility networks needs to be closely coordinated with urban development. In many periurban neighborhoods, expanding utility networks is hampered by the absence of legal tenure and high rates of tenancy, not to mention inadequate spacing of dwellings. Providing services to these communities will require close cooperation with urban authorities, because many of these issues can be resolved only if they are addressed in a synchronized and coordinated manner.

Take a Hard-headed Look at Affordability
Underrecovery of costs has serious implications for the financial health of utilities and slows the pace of service expansion. Many of Africa's power utilities capture only two-thirds of the revenue they need to function sustainably. This revenue shortfall is rarely covered through timely and explicit fiscal transfers. Instead, maintenance and investment activities are cut back to make ends meet, which starves the utility of funds to expand service coverage and cuts the quality of service to existing customers.

Affordability, the usual pretext for underpricing services, does not bear much scrutiny. The political economy likely provides the real explanation for low tariffs: Populations currently connected to utility services tend to be those with the greatest voice. The implicit subsidies created by underpricing are extremely pro-rich in their distributional incidence. In all but the poorest African countries, service coverage could be substantially increased before any real affordability problems would be encountered. In the poorest of the low-income countries, affordability is a legitimate concern for the bulk of the population and would constrain universal coverage. Even in the poorest countries, however, recovering operating costs should be feasible, with subsidies limited to capital costs.

What effect would removing utility subsidies have on reducing poverty? For most countries, electricity spending accounts for only a tiny fraction of total consumption. At the national level, the impact of a 50 percent increase in tariffs or even of a doubling of tariffs is marginal; the share of the population living in poverty increases barely one-tenth of a percentage point. Among households with a connection to the network, the impact is larger but still limited. Indeed, rarely is there more than a one or two percentage point increase in the share of households in poverty. Because the households that benefit from a connection tend to be richer than other households, the increase in poverty starts from a low base. So the small impact of an increase in tariffs on poverty could be offset by reallocating utility subsidies to other areas of public expenditure with a stronger pro-poor incidence.

Tariff increases can be either phased in gradually or effected instantly through a one-time adjustment. Both approaches have advantages and disadvantages. The public acceptability of tariff increases can be enhanced if they form part of a wider package of measures that includes service quality improvements. One way to strengthen social accountability is to have communication strategies link tariffs with service delivery standards and suggest conservation measures to contain the overall bills. Either way, it is perhaps most important to ensure that the realignment of tariffs and costs is not temporary by providing for automatic indexing and periodic revisions of tariffs.

In the absence of a strong payment culture, customers who object to tariff hikes may refuse to pay their bills. Therefore, even before addressing tariff adjustments, it is important for utilities to work on raising revenue collection rates toward best practice levels and establishing a payment culture. At least for power, one technological solution is to use prepayment meters, which place customers on a debit card system similar to that used for cellular telephones. For utilities, this eliminates credit risk and avoids nonpayment. For customers, this allows them to control their expenditure and avoid consuming beyond their means. South Africa was at the forefront in development of the keypad-based prepayment electricity meter with the first product, called Cashpower, launched by Spescom in 1990. Tshwane, also in South Africa, reports universal coverage of its low-income consumers with prepayment meters. In Lesotho, Namibia, and Rwanda, a majority of residential customers are on prepayment meters. In Ghana and Malawi, a clear policy has been pursued of rapidly increasing the share of residential customers on prepayment meters (figure 5.11).

Target Subsidies to Promote Service Expansion

Subsidies have a valuable and legitimate role in the right circumstances. They may be appropriate when households genuinely cannot purchase a subsistence allowance of a service that brings major social and economic benefits to them and those around them, as long as governments can afford to pay those subsidies. However, utility subsidies' design and targeting needs to be radically improved to fulfill their intended role.

Figure 5.11   Prepayment Metering

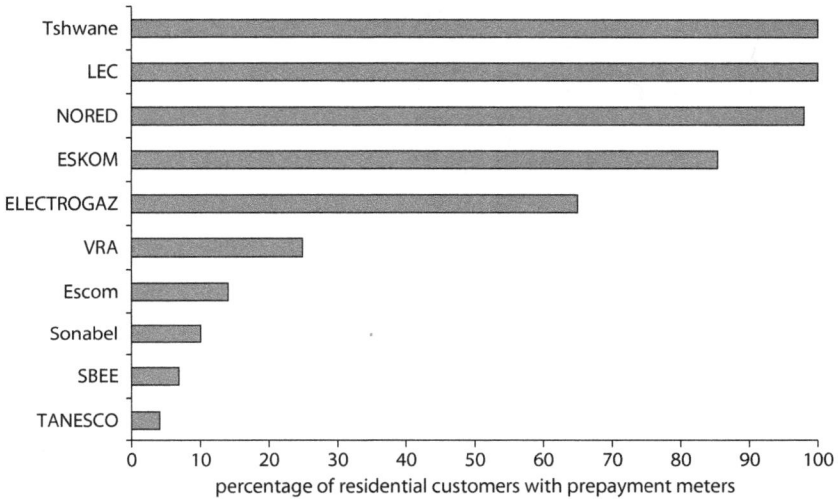

percentage of residential customers with prepayment meters

Source: Foster and Briceño-Garmendia 2009.

As noted previously, the utility subsidies practiced in Africa today largely bypass the poor.

African utilities typically subsidize consumption, but subsidizing connection is potentially more equitable and effective in expanding coverage. The affordability problems associated with connection charges are often much more serious than those associated with use-of-service charges. Given that connections are disproportionately concentrated among the more affluent, the absence of a connection is disproportionately concentrated among the poorest. This could make the absence of a connection a good targeting variable.

Where coverage is far from universal even among the higher-income groups, who will likely be the first to benefit from coverage expansion, connection subsidies may be just as pro-rich as consumption subsidies. Simulations suggest that the share of connection subsidies going to the poor would be only about 37 percent of the share of the poor in the population; this is a highly pro-rich result no better than that of existing consumption subsidies (table 5.3).

Limiting subsidies to connections in new network rollout as opposed to densification of the existing network would substantially improve targeting. The share of connection subsidies going to the poor would rise to

Table 5.3   Potential Targeting Performance of Electricity Connection Subsidies under Various Scenarios

Scenarios	Targeting performance
1. New connections mirror pattern of existing connections	0.37
2. Only households beyond reach of existing network receive connection subsidies	0.95
3. All unconnected households receive subsidy	1.18

Source: Banerjee and others 2008; Wodon 2007a, b.
Note: A measure of distributional incidence captures the share of subsidies received by the poor divided by the proportion of the population in poverty. A value greater than one implies that the subsidy distribution is progressive (pro-poor), because the share of benefits allocated to the poor is larger than their share in the total population. A value less than one implies that the subsidy distribution is regressive (pro-rich).

95 percent of their share in the population, but the outcome would remain pro-rich. Providing a connection subsidy equally likely to reach all unconnected households would ensure that the percentage going to the poor exceeds their share of the population by 118 percent. This strategy ultimately achieves a progressive result. To improve the distributional incidence beyond this modest level would require connection subsidies to be accompanied by other socioeconomic screens. In the low-access environment in most African countries, the absence of a connection remains a fairly weak targeting variable.

Can anything be done to improve the impact of use-of-service subsidies? The poor performance of existing utility subsidies is explained partly by pro-rich coverage but also by the widespread use of poorly designed IBTs. Common design failures in power IBTs include large subsistence thresholds, so that only consumers with exceptionally high consumption contribute fully to cost recovery (Briceño-Garmendia and Shkaratan 2010). Some improvements in targeting could be achieved by eliminating fixed charges, reducing the size of first blocks to cover only genuinely subsistence consumption, and changing from an IBT to a volume-differentiated tariff where those consuming beyond a certain level forfeit the subsidized first block tariff completely. Even with these modifications, however, the targeting of such tariffs would improve only marginally and would not become strongly pro-poor in absolute terms.

Global experience suggests that utility subsidy targeting can be improved and become reasonably progressive if some form of geographical or socioeconomic targeting variables can be used beyond the level of consumption (Komives and others 2005). Such targeting schemes hinge, however, on the existence of household registers or property cadastres that support the classification of beneficiaries, as well as a significant amount of administrative capacity. Both factors are often absent in Africa, particularly in the low-income countries.

Utility service underpricing that benefits just a small minority of the population costs many African countries as much as 1 percent of GDP. As countries move toward universal access, that subsidy burden would increase proportionately and rapidly become unaffordable for the national budget. So countries should consider how the cost of any proposed subsidy policy would escalate as coverage improves. This test of a subsidy's fiscal affordability is an important reality check that can help countries avoid embarking on policies that are simply not scalable.

One other potentially effective method of targeting is to limit the allocation of subsidies to lower-cost and lower-quality alternatives that encourage self-selection, such as load-limited supplies. The theory is that more affluent customers will eschew second-best services and automatically select to pay the full cost of the best alternative, thus identifying themselves and leaving the subsidized service to less affluent customers.

Systematic Planning Is Needed for Periurban and Rural Electrification

As already noted, the majority of the population in Sub-Saharan Africa still resides in rural areas. Some countries have a much higher potential for making rural electrification advances more cost effective, because a higher proportion of their population lives close to existing networks (figure 5.12). Thus Benin, Ghana, Lesotho, Rwanda, Senegal, and Uganda are more favorably positioned than, for example, Burkina Faso, Chad, Madagascar, Mozambique, Niger, Tanzania, or Zambia.

The potential for extending access in a given situation depends on population density, distance from the grid, economic activity, and developmental needs. Because those circumstances differ widely across regions and countries, the most successful rural electrification will be selective, detailed, and carefully planned. Data show that those countries with clear planning criteria have generally been more successful at rural electrification.

Given the scale of investments needed, a systematic approach to planning and financing new investments is critical. The current project-by-project, ad hoc approach in development partner financing has led to fragmented planning, volatile and uncertain financial flows, and duplication of efforts. Engagement across the sector in multiyear programs of access rollout supported by multiple development partners as part of a

Figure 5.12   Potential Rural Access: Distribution of Population by Distance from Substation

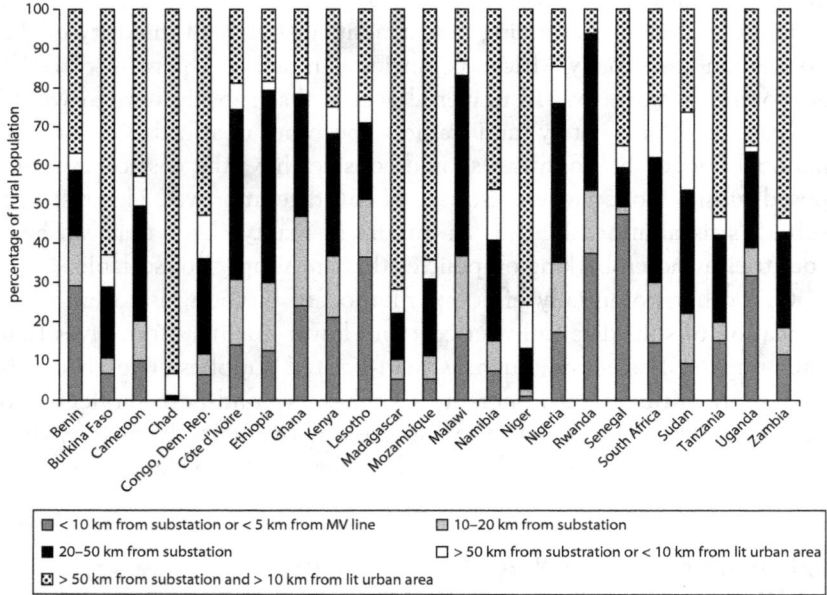

Source: Eberhard and others 2008.
Note: Transmission lines are not available for Chad or Niger, so "remote" potential service area is overestimated.

coherent national strategy will channel resources in a more sustained and cost-effective way to the distribution subsector. Coordinated action by development partners will also reduce the unit costs of increasing access by achieving economies of scale in implementation.

Countries with dedicated REFs have achieved higher rates of electrification than those without. Of greatest interest, however, are the differences among the countries that have funds. Case studies indicate that the countries that have taken a centralized approach to electrification—with the national utility made responsible for extending the grid—have been more successful than those that followed decentralized approaches. Undoubtedly, those REAs that have attempted to recruit multiple utilities or private companies into the electrification campaign have a contribution to make (see box 5.3), especially in promoting minigrids and off-grid options. These should be seen, however, as complementary to the main utility's efforts to extend the grid.

Box 5.3

Rural Electrification in Mali

Among new rural electrification agencies created in Africa, Mali's AMADER (Agence Malienne pour le Developpement de l'Energie Domestique et d'Electrification Rurale) has had considerable success. In Mali, only 13 percent of the rural population has access to electricity. Until they are connected, most rural households meet their lighting and small power needs with kerosene, dry cell, and car batteries, with an average household expenditure of $4–$10 per month. About half of Mali's 12,000 villages have a school or health center clinic or both; however, most are without any form of energy for lighting or for operating equipment. The majority of Malians—more than 80 percent—use wood or charcoal for cooking and heating. The use of these sources of energy make the poor pay about $1.50 per kWh for energy, more than 10 times the price of a kilowatt-hour from the grid. In addition to rural electrification, AMADER promotes community-based woodland management to ensure sustainable wood fuel supply. It also has interfuel substitution initiatives and programs for the introduction of improved stoves.

AMADER, created by law in 2003, employs two major approaches to rural electrification: spontaneous "bottom-up" electrification of specific communities and planned "top-down" electrification of large geographic areas. To date, the bottom-up approach, which typically consists of minigrids operated by small local private operators, has been more successful. Eighty electrification subprojects managed by 46 operators are financed so far through the bottom-up approach. By late December 2009, connections had been made to more than 41,472 households, 803 community institutions, 172 schools, and 139 health clinics. Typically, AMADER provides grants for 75 percent of the start-up capital costs of rural electrification subprojects, depending on the proposed connection target within the first two years, the average cost per connection, and the average tariff.

Most of the bottom-up rural electrification subprojects are based on conventional, diesel-fueled minigrids with installed generation capacities mainly below 20 kilowatts. Customers on these isolated minigrids typically receive electricity for six to eight hours daily. In promoting these new projects, AMADER performs three main functions. It is a provider of grants, a supplier of engineering and commercial technical assistance, and a de facto regulator through its grant agreements with operators. The grant agreement can be viewed as a form of "regulation by contract," because it establishes minimum standards for technical and commercial quality of service and maximum tariffs allowed for both metered and unmetered customers.

(continued next page)

Box 5.3 *(continued)*

Renewable energy technologies, particularly solar photovoltaics, have been successfully introduced into Mali's rural energy mix. Over a period of six years, more than 7,926 solar home systems and more than 500 institutional solar photovoltaic systems were installed countrywide. A solar power station of 72 kW peak solar photovoltaic plant connected to an 8 kilometer distribution network in the village of Kimparana, the first of its type and scale in West Africa, has been operational since 2006. It is providing power to about 500 households, community institutions, and microenterprises. Biofuels are also being promoted for electricity production in the village of Garalo in partnership with the Mali Folkecenter, a local nongovernmental organization (NGO).

Women's associations are also playing an important role in remote communities as providers of energy services. They manage some of the multifunctional platforms after receiving training in basic accounting in local languages provided by NGOs financed through the project. To date, multifunctional platforms have been installed in 64 communities and have resulted in 7,200 connections. A multifunctional platform is composed of a small 10 kW diesel engine coupled to a generator. The platform can be connected to income-generating equipment, such as cereal grinding mills, battery chargers, dehuskers, and water pumps. AMADER has added public lighting networks of about 2 kilometers to the multifunctional platforms in about 35 communities.

To ensure that the projects are financially sustainable, AMADER permits operators to charge residential and commercial cost-reflective tariffs that are often higher than the comparable tariffs charged to grid-connected customers. For example, the energy charge for metered residential customers on isolated minigrids is about 50 percent higher than the comparable energy charge for grid-connected residential customers served by EDM (Electricidade de Moçambique, the national electric utility). Many of the minigrid operators also provide service to unmetered customers. Unmetered customers are usually billed on a flat monthly charge per light bulb and power outlet, combined with load-limiting devices, to ensure that a customer does not connect appliances above and beyond what he or she has paid for.

To reduce financial barriers for operators, leasing arrangements have been proposed, as well as a loan guarantee program for Malian banks and microfinance institutions that would be willing to provide loans to potential operators and newly connected customers to increase productive energy uses. Work is ongoing to attract private operators to larger concessions and to increase the share of renewable energies in Mali's rural energy mix.

Source: Interviews with World Bank staff from the Africa Energy Department, 2008.

In an African context, it is legitimate to ask how far it is possible to make progress with rural electrification when the urban electrification process is still far from complete. Across countries, a strong correlation is found between urban and rural electrification rates, as well as a systematic lag between the two. Countries with seriously underdeveloped generation capacity and tiny urban customer bases are not well placed to tackle the challenges of rural electrification, either technically because of power shortages or financially because of the lack of a basis for cross-subsidization. Dedicated electrification funds should thus also be made available for periurban connections.

It is also important to find ways to spread the benefits of electrification more widely, because universal household electrification is still decades away in many countries. Sectorwide programmatic approaches must ensure that the benefits of electrification touch even the poorest households that are too far from the grid or unable to pay for a grid connection. Street lighting may be one way to do this in urban areas. In rural areas, solar-powered electrification of clinics and schools that provide essential public services to low-income communities is one way to allow them to participate in the benefits of electrification. Another way is appropriate technology, such as low-cost portable solar lanterns that are much more accessible and affordable to the rural public. The "Lighting Africa" initiative is supporting the development of the market for such products.

Finally, the difficult question needs to be posed as to whether aggressive electrification will exacerbate the financial problems of the sector. Diverting scarce capital to network expansion can easily result in a familiar situation where investments barely generate adequate revenue to support operating and maintenance costs, with no contribution to refurbishment or capital-replacement requirements. The resulting cash drains on the utility could be serious. Ultimately, difficult choices need to be made on how to allocate scarce capital. Should it go to network expansion, or are investments in new generation capacity more important? In either case, careful tradeoffs will be required.

References

Banerjee, Sudeshna, Quentin Wodon, Amadou Diallo, Taras Pushak, Hellal Uddin, Clarence Tsimpo, and Vivien Foster. 2008. "Access, Affordability and Alternatives: Modern Infrastructure Services in Sub-Saharan Africa." Background Paper 2, Africa Infrastructure Country Diagnostic, World Bank, Washington, DC.

Boccanfuso, Dorothée, Antonio Estache, and Luc Savard. 2009. "Distributional Impact of Developed Countries' CC Policies on Senegal: A Macro-Micro CGE Application." Cahiers de Recherche 09-11, Department of Economics, Faculty of Administration, University of Sherbrooke, Quebec, Canada.

Boccanfuso, Dorothée, and Luc Savard. 2000. "The Food Crisis and Its Impact on Poverty in Senegal and Mali: Crossed Destinies." GREDI, Working Paper 08-20, University of Sherbrook, Quebec, Canada.

———. 2005. "Impact Analysis of the Liberalization of Groundnut Production in Senegal: A Multi-Household Computable General Equilibrium Model." Cahiers de Recherche 05-12, Department of Economics, Faculty of Administration, University of Sherbrooke, Quebec, Canada.

Briceño-Garmendia, Cecilia, and Maria Shkaratan. 2010. "Power Tariffs: Caught between Cost Recovery and Affordability." Working Paper 20, Africa Infrastructure Country Diagnostic, World Bank, Washington, DC.

Clark, Alix, Mark Davis, Anton Eberhard, and Njeri Wamakonya. 2005. "Power Sector Reform in Africa: Assessing the Impact on Poor People." ESMAP Report 306/05, World Bank, Washington, DC.

Eberhard, Anton, Vivien Foster, Cecilia Briceño-Garmendia, Fatimata Ouedraogo, Daniel Camos, and Maria Shkaratan. 2008. "Underpowered: The State of the Power Sector in Sub-Saharan Africa." Background Paper 6, Africa Infrastructure Sector Diagnostic, World Bank, Washington, DC.

ESMAP (Energy Sector Management Assistance Program). 2007. "Technical and Economic Assessment of Off-Grid, Mini-Grid and Grid Electrification Technologies." ESMAP Technical Paper 121/07, World Bank, Washington, DC.

Foster, Vivien, and Cecilia Briceño-Garmendia, eds. 2009. *Africa's Infrastructure: A Time for Transformation.* Paris, France, and Washington, DC: Agence Française de Développement and World Bank.

Gulyani, S., D. Talukdar, and D. Jack. 2008. "A Tale of Three Cities: Understanding Differences in Provision of Modern Services." Working Paper 10, African Infrastructure Country Diagnostic, World Bank, Washington, DC.

Komives, Kristin, Vivien Foster, Jonathan Halpern, and Quentin Wodon. 2005. *Water, Electricity, and the Poor: Who Benefits from Utility Subsidies?* Washington, DC: World Bank.

Marquard, A., B. Bekker, A. Eberhard, and T. Gaunt. 2008. "South Africa's Electrification Programme: An Overview and Assessment." *Energy Policy* 36: 3125–37.

Mostert, W. 2008. "Review of Experience with Rural Electrification Agencies: Lessons for Africa." EU Energy Initiative Partnership Dialogue Facility, Eschborn, Germany.

Wodon, Quentin, ed. 2007a. "Electricity Tariffs and the Poor: Case Studies from Sub-Saharan Africa." Working Paper 11, Africa Infrastructure Country Diagnostic, World Bank, Washington, DC.

———. 2007b. "Water Tariffs, Alternative Service Providers, and the Poor: Case Studies from Sub-Saharan Africa." Working Paper 12, Africa Infrastructure Country Diagnostic, World Bank, Washington, DC.

CHAPTER 6

Recommitting to the Reform of State-Owned Enterprises

Most electricity utilities in Sub-Saharan Africa are state owned. Yet most of them are inefficient and incur significant technical and commercial losses. Hidden costs abound in the sector: network energy losses, underpricing, poor billing and collections practices resulting in nonpayment and theft, and overstaffing all absorb revenue that could be used for maintenance and system expansion.

Evidence suggests that reforms in the governance of state-owned enterprises (SOEs) could reduce hidden costs. This has even happened in some African countries. Data gathered by the Africa Infrastructure Country Diagnostic show that those enterprises that have implemented more governance reforms have benefited from improved performance.

No single reform will be sufficient to effect lasting improvements in performance. Rather, an integrated approach to governance reform is needed. Roles and responsibilities need to be clarified, which will involve clear identification, separation, and management of government's different roles in policy making, ownership of utility assets, and regulation. Roles and responsibilities can further be clarified through public entity legislation, corporatization, codes of corporate governance, performance contracts, effective supervisory and monitoring agencies, and transparent transfers for social programs.

Another broad set of reforms involves strengthening the role of interest groups with a stake in more commercial behavior—for example, taxpayers, customers, and private investors. This can be promoted through direct competition, improved transparency and information, and commercialization practices such as outsourcing, mixed-capital enterprises, and structural reform.

Hidden Costs in Underperforming State-Owned Enterprises

The previous chapters have highlighted the deficits of the power sector in Africa. Not only is there insufficient generating capacity, but also national utilities have performed poorly both financially and technically.

Average distribution losses in Africa are 23 percent compared with the commonly used norm of 10 percent or less in developed countries. Moreover, average collection rates are only 88.4 percent compared with best practice of 100 percent.

Underpricing and inefficiency generate substantial hidden costs for the region's economy. Combining the costs of distribution losses and uncollected revenue and expressing them as a percentage of utility turnover provides a measure of the inefficiency of utilities (figure 6.1). The inefficiency of the median utility is equivalent to 50 percent of turnover, which means that only two-thirds of revenue is captured. The inefficiency of the utilities creates a fiscal drain on the economy, because governments must frequently cover any operating deficit to prevent the utility from becoming insolvent.

Inefficiencies also seriously undermine the utilities' future performance. Utility managers with operating deficits are often forced to forgo maintenance. Inefficient operation has a similar adverse effect on investment. For example, countries with below-average efficiency have increased electrification rates by only 0.8 percent each year, compared with 1.4 percent for utilities with above-average efficiency. Less efficient utilities also have greater difficulty in meeting demand for power. In countries with utilities of below-average efficiency, suppressed or unmet power demand accounts for 12 percent of total demand, compared with only 6 percent in countries with utilities of above-average efficiency (figure 6.2).

Chapter 7 explores more quantitative measures of inefficiency and hidden costs and their effect on funding requirements.

Figure 6.1   Overall Magnitude of Utility Inefficiencies as a Percentage of Revenue

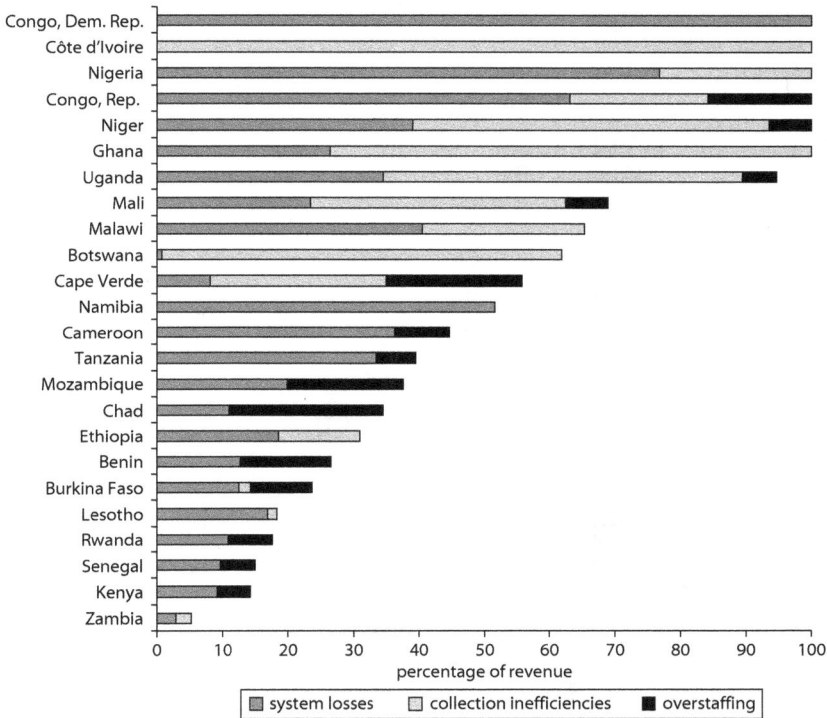

Source: Briceño-Garmendia and Shkaratan 2010.

Driving Down Operational Inefficiencies and Hidden Costs

Countries that have made progress in power sector reform, including regulatory reform, have substantially lower hidden costs (figure 6.3). In particular, private sector participation and the adoption of contracts with performance incentives by state-owned utilities appear to substantially reduce hidden costs. The case of Kenya Power and Lighting Company (KPLC) is particularly striking (box 6.1).

Over the years, countries have spent substantial sums on institutional reforms in the power sector, including management training, improved internal accounting and external auditing, improved boards of directors, financial and operational information and reporting systems, and establishment and strengthening of supervisory and regulatory agencies. Some

Figure 6.2   Effect of Utility Inefficiency on Electrification and Suppressed Demand

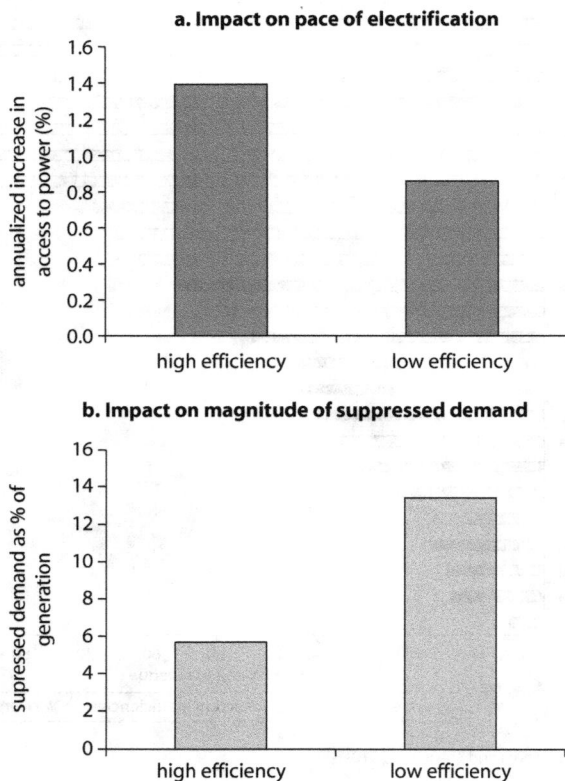

a. Impact on pace of electrification

b. Impact on magnitude of suppressed demand

Source: Derived from Eberhard and others 2008.

successes have endured (see box 6.2), but in many other cases reforms have not had the intended effect.

Effect of Better Governance on Performance of State-Owned Utilities

Evidence is increasing that governance reform can improve the perform-ance of state-owned utilities. Governance may be assessed using various criteria, including ownership and shareholder quality, managerial and board autonomy, accounting standards, performance monitoring, out-sourcing to the private sector, exposure to labor markets, and the disci-pline of capital markets (Vagliasindi 2008).

Figure 6.3   Impact of Reform on Hidden Costs in the Power Sector in Sub-Saharan Africa

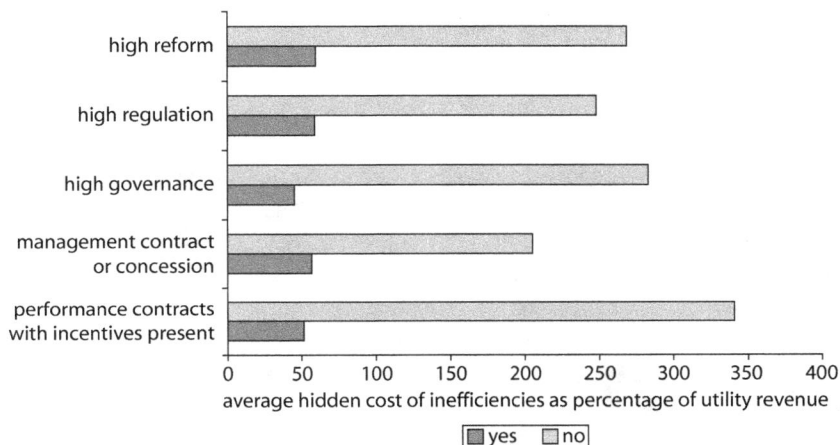

average hidden cost of inefficiencies as percentage of utility revenue

☐ yes   ☐ no

Source: Eberhard and others 2008.

Good governance is not universal among Sub-Saharan Africa utilities (figure 6.4). The most prevalent good governance practices are those relating to managerial autonomy. Most utilities report requirements to be profitable and pay market rates for debt, but the vast majority benefit from sizeable subsidies and tax breaks and are not financially sound enough to borrow. Only 60 percent of the sample utilities publish audited accounts, and stock exchange listing is virtually unheard of (Kengen and KPLC in Kenya are the exceptions). Overall, most utilities in the sample meet only about half of the criteria for good governance.

A comparison of utilities based on 35 governance indicators provides striking and consistent evidence that good governance improves utility performance (figure 6.5).

Making State-Owned Enterprises More Effective

Two broad sets of governance reforms are important to ensure that improvements to the performance of state-owned utilities are sustainable. First, roles and responsibilities need to be clarified. This involves clear identification, separation, and management of government's different roles in policy making, ownership of utility assets, and regulation of prices and quality of utility services. Roles and responsibilities can further be

Box 6.1

Kenya's Success in Driving Down Hidden Costs

In the early 2000s, hidden costs in the form of underpricing, collection losses, and distribution losses on the part of Kenya's power distribution utility (KPLC) absorbed as much as 1.4 percent of Kenya's gross domestic product (GDP) per year. Management reforms resulted in revenue collection improvement—from 81 percent in 2004 to 100 percent in 2006. Distribution losses also began to fall, though more gradually, which reflected the greater technical difficulty they posed. Power-pricing reforms also allowed tariffs to rise in line with escalating costs from $0.07 in 2000 to $0.15 in 2006 and $0.20 in 2008. As a result of reforms, hidden costs in Kenya's power sector fell to 0.4 percent of GDP by 2006 and almost to zero by 2008 (see figure), among the lowest totals of any African country.

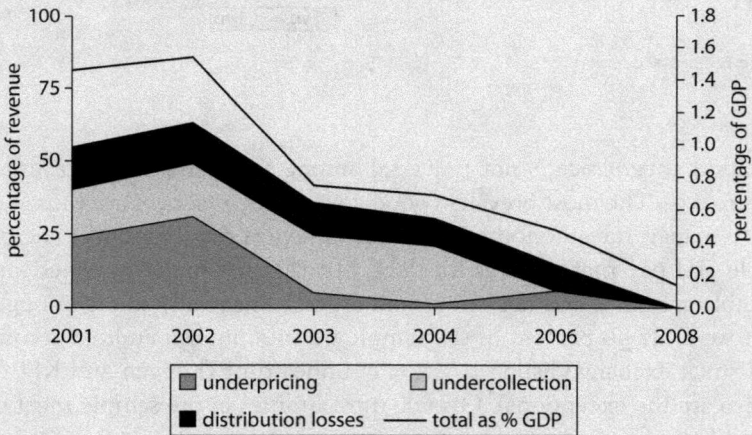

Source: Foster and Briceño-Garmendia 2009.
Note: GDP = gross domestic product.

clarified through public entity legislation, corporatization, codes of corporate governance, performance contracts, effective supervisory and monitoring agencies, and transparent transfers for social programs.

The second broad set of reforms revolves around what Gomez-Ibanez (2007, 33–48) refers to as "changing the political-economy of an SOE," by which he means strengthening the role of other power-sector stakeholders, such as taxpayers, customers, and private investors. This can be promoted through improved transparency, commercialization

Box 6.2

Botswana's Success with a State-Owned Power Utility

The state-owned electricity utility Botswana Power Corporation (BPC) was formed by government decree in 1970 to expand and develop electrical power potential in the country. The utility began as one power station in Gaborone with a network that extended about 45 kilometers outside the city. Since then, its responsibilities and the national network have expanded enormously. The government regulates the utility through the Energy Affairs Division of the Ministry of Minerals, Energy and Water Affairs.

During the tenure of BPC, access to electricity increased to 22 percent in 2006 and is set to reach 100 percent by 2016. Government funding has allowed BPC to extend the electricity grid into rural areas. The power system is efficient, with distribution losses of less than 10 percent and a return on assets equal to its cost of capital.

When capacity shortages seem likely, BPC must decide between importing power and expanding its own generation facilities. The national system, in 2005, provided 132 megawatts, and neighboring countries supplied another 266 megawatts via the Southern African Power Pool; Botswana has been an active member and major beneficiary of the regional pool since its inception in 1995. Its active trading position has helped to promote multilateral agreements and enhance cooperation among pool members.

To be fair, BPC has benefited from the availability of cheap imported power from South Africa (which is now severely threatened by a power crisis there). Regardless, analysts contend that BPC's strong performance is equally attributable to institutional factors: a strong, stable economy, cost-reflective tariffs, lack of government interference in managerial decisions, good internal governance, and competent and motivated employees.

Source: Molefhi and Grobler 2006.

practices, structural reform, direct competition, and mixed-capital enterprises (table 6.1).

Defined Roles and Responsibilities
Utilities management in Sub-Saharan Africa often suffers from mixed— and sometimes contradictory—policy and governance directives and incentives. Governments can interfere with management decisions in an

Figure 6.4   Incidence of Good-Governance Characteristics among State-Owned Utilities

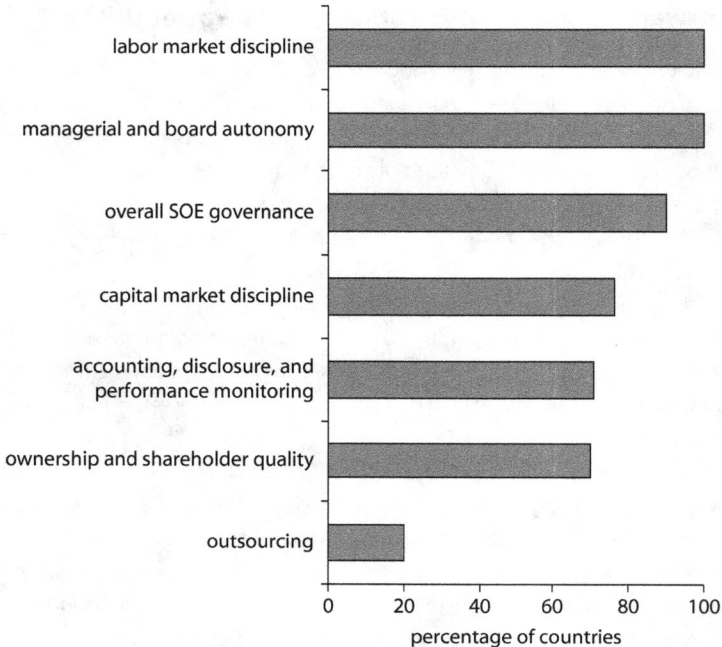

Source: Eberhard and others 2008.
Note: SOE = state-owned enterprise.

ad hoc and nontransparent manner in areas such as overstaffing and excessive salary levels. They may also pressure utilities to electrify certain areas, ignore illegal connections and nonpayment, or maintain excessively low prices. Government may also be unclear about its role as owner of the utility and the need to maintain and expand its assets. Regulation of prices and quality of service may also be arbitrary and unconnected to ensuring the financial sustainability of the utility. The combination of these nontransparent and sometimes contradictory pressures on the management of the utility can be disastrous. Inevitably investment is insufficient, and service quality deteriorates.

These challenges can be addressed by clearly identifying, separating, and coordinating government's different roles and functions in the sector. Clear policy statements can help clarify and make transparent government's social, economic, and environmental objectives. Sector and public entity

Figure 6.5   Effect of Governance on Utility Performance in State-Owned Power Utilities

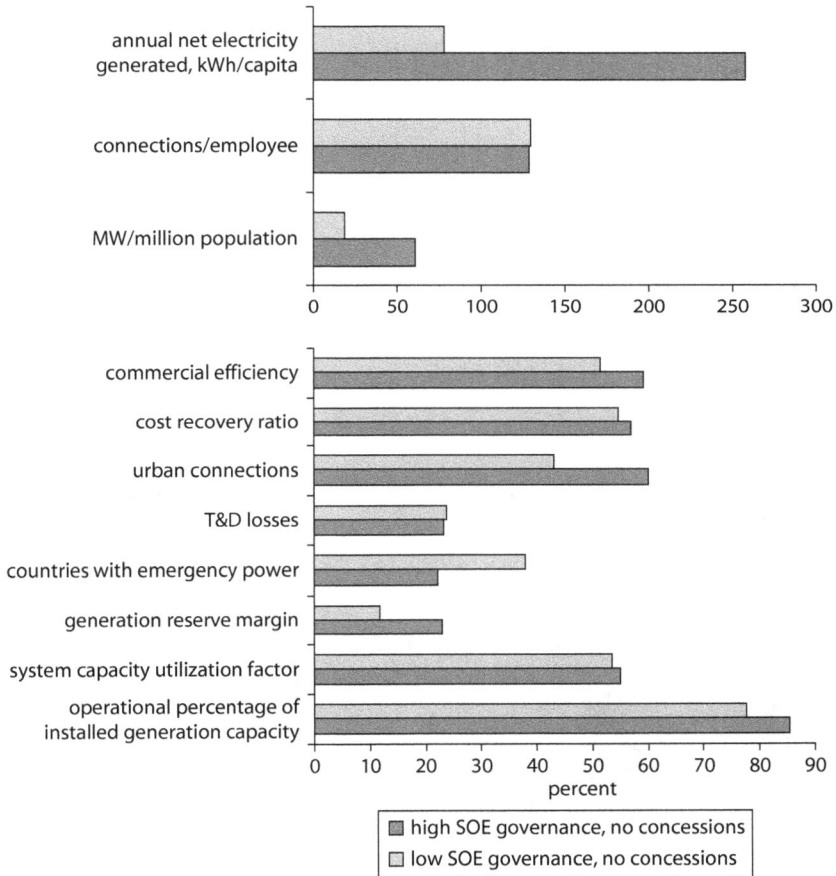

Source: Eberhard and others 2008.
Note: kWh = kilowatt-hour; MW = megawatt; SOE = state-owned enterprise; T&D = transmission and distribution.

legislation can also clarify and separate a government's policy role from its shareholding function and the necessity of balancing demands for more affordable electricity tariffs with the necessity of maintaining financial sustainability (box 6.3). It makes sense to separate the policy-making ministry from the SOE shareholding ministry so that they focus clearly on their respective mandates. However, effective policy coordination will also be needed at the cabinet level to achieve the necessary tradeoffs between social and economic objectives.

Table 6.1   Governance Reforms to Improve State-Owned Utility Performance

Clarification of roles and responsibilities	Changing the political economy of the utility
• Identification, separation, and coordination of government's different roles as policy maker, asset owner, and regulator • Public entity legislation • Corporatization • Codes of corporate governance • Performance contracts • Effective supervisory or monitoring agencies • Transparent transfers for social programs • Independent regulator	• Improved transparency and information • Commercialization and outsourcing • Labor market reform • Structural reform and direct competition • Mixed-capital enterprises • Customer-owned enterprises

Source: Eberhard and others 2008.

Box 6.3

The Combination of Governance Reforms That Improved Eskom's Performance

The experience of Eskom, South Africa's national electricity utility, provides a model for the implementation of governance reforms. A clear distinction is now made between the shareholder ministry (Public Enterprises) and the sector policy ministry (Energy). In addition, an independent authority regulates market entry through licenses, sets tariffs, and establishes and monitors technical performance and customers' service standards. Eskom was corporatized through the Eskom Conversion Act and is subject to ordinary corporate law. It must pay dividends and taxes and publish annual financial statements according to international accounting standards. The board (appointed by the Minister of Public Enterprises) is responsible for day-to-day management subject to a performance contract that includes a range of key performance indicators.

Additional legislation (the Public Finance Management Act and the Promotion of Administrative Justice Act) defines in more detail how the utility should handle finance, information disclosure, reporting, and authorizations. A general corporate governance code also applies to all state-owned enterprises. The performance contract is monitored, albeit not very effectively, by the Ministry of Public Enterprises. The utility benefits from separate subsidies for electrification connections and for consumption (poor households receive their first 50 kilowatt-hours each month free of charge).

(continued next page)

Box 6.3 *(continued)*

After reforms in the 1980s and the appointment of an experienced private sector manager as Eskom's chief executive officer, a commercial culture was embedded within the utility, separate business units were created with business plans and new budgeting and accounting systems, and outsourcing was used more widely.

Eskom is a mixed-capital enterprise. Although wholly owned by the state, it raises capital on private debt markets, locally and internationally, through issuing bonds. It is rated by all the major global credit agencies. Eskom managers are acutely aware that their financial performance is subject to thorough external scrutiny. Any possible downgrading of their debt can make capital scarce or more expensive when they embark on a major capital expansion program.

These reforms have caused Eskom to perform relatively well compared with other African utilities. Recently, however, Eskom has had to institute load shedding because it has had insufficient generation capacity to meet demand. Policy uncertainties and an earlier prohibition on Eskom's investing in new capacity while private sector participation was being considered have led to capacity shortages.

What Eskom lacks most of all is direct competition. Eskom is dominant in the region; it generates 96 percent of South Africa's electricity, transmits 100 percent, and distributes approximately 60 percent. Neither government nor the regulator has a good enough idea of Eskom's actual efficiency or inefficiency. Indications suggest that planning and cost controls could improve. Only direct competitors could provide an appropriate benchmark.

Source: Authors.

An independent regulatory authority is better positioned to balance the need for protecting consumers (price and quality of service) with providing incentives for utilities to reach financial sustainability by reducing costs, improving efficiency, and moving toward more cost-reflective pricing.

Corporatization of state-owned utilities further helps clarify government's role as owner and shareholder. Typically the utility will be made subject to ordinary company law. Government is the shareholder, but the utility has a legal identity that is separate from government. The board also includes independent and nonexecutive directors with legal rights and obligations, which makes political interference more difficult. Corporatized utilities have separate accounts and are typically liable for paying taxes and dividends.

Legislation that brings about corporatization also clarifies the mandate, powers, and duties of the utility and its board, the utility's obligation to earn a profit or an adequate return on assets, and its financing and borrowing permissions. Responsibilities for financial management, budgeting processes, accounting, reporting, and auditing are also clearly defined. Codes of corporate governance may also be adopted to clarify and define the relationship between the shareholder and the utility's board as well as the way in which the board and management operate.

A shareholder compact or performance contract usually sets out the shareholder ministry's objectives for the utility. It specifies the obligations and responsibilities of the enterprise, on the one hand, and the "owner" (that is, the ministry, the supervisory body, or the regulator), on the other. Performance contracts are negotiated, written agreements that clarify objectives of governments and motivate managers to achieve improved performance. They normally address tariffs, investments, subsidies, and noncommercial (social or political) objectives and their funding; they sometimes include rewards for good managerial (and staff) performance and, more rarely, sanctions for nonfulfillment of objectives.

Performance contracts are also used to reveal information and to monitor managers' performance. Typically they include elements of business plans and specify a number of key performance measures and indicators. Performance contracts can also be used between central SOE boards and decentralized units. Performance indicators could include the following: net income, return on assets, debt and equity ratios, interest cover, dividend policy, productivity improvements, customer satisfaction indexes, connection targets, human resource issues, procurement policy, and environmental adherence.

Performance contracts are widespread, but their effectiveness is not guaranteed. They have not always reduced the information advantage that managers enjoy over owners, which often allows managers to negotiate performance targets that are easy for the utility to achieve. Furthermore, managers are not convinced of the credibility of government promises, and they have not been sufficiently motivated by rewards and penalties. This is understandable, considering that contracts often lack mechanisms for enforcing government commitments to pay utility bills or penalize underperforming managers.

At the heart of the challenge of making performance contracts work more effectively are the classic principal-agent and moral hazard problems. Politicians may not benefit from better performance and may subsequently try to make managers serve objectives that conflict with efficiency, such as rewarding political supporters with jobs or subsidies. Contracts can also be incomplete and fail to anticipate events and contingencies. Finally, governments can renege on commitments, including promised budgets for social programs. Performance contracts are therefore not a panacea and should be used only if governments are prepared to deal with the challenges of information asymmetry, effective incentives, and credible commitments.

In the end, the extent of hidden costs and inefficiencies that affect African utilities is not accurately known. Basic operational and financial data on firm performance are either not collected, not sent to supervisors, not tabulated and published by the supervising bodies, or not acted upon. In the absence of information—or of action taken on the basis of what information is produced—improved performance cannot be expected. Independent supervisory units that can effectively monitor performance contracts are therefore essential. They would preferably be located in the Ministry of Finance or in a dedicated Public Enterprises Ministry. The policy or sector ministry may be hindered by a focus on short-term social or political outcomes rather than on efficiency and financial sustainability. Alternatively, the supervisory function could be contracted out to an expert panel.

Other reforms could include hiring private sector managers to instill a commercial culture in the utility. This would ensure that tariffs are high enough to provide sufficient revenue, the utility earns a rate of return at least equal to its cost of capital, billing and collection approaches 100 percent, and customer service improves. The reforms will eliminate government subsidies of the utility's cost of capital. Instead, the utility will be required to raise finance from private capital markets. Employment and procurement should be undertaken on a commercial basis, and utilities will be encouraged to outsource functions that another company can perform more efficiently. Competition among suppliers for outsourcing contracts could also drive costs down.

Finally, commercial responsibilities should be clearly separated from social goals by establishing transparent mechanisms such as fiscal transfers and subsidies for connections for poor households. This would allow utility managers to focus on improving operational efficiency.

Altering the Political Economy around the Utility
Governance reforms should also strengthen other stakeholders with an interest in reduced operating losses and improved operating performance. These reforms could encompass improved transparency and flow

of information, including comprehensive annual reports and financial statements, performance contracts (made available publicly along with results), investment and coverage plans, prices, costs and tariffs, service standards, benchmarking, and customer surveys. Information needs to be credible, coherent, and timely. However, better dissemination of information alone is not sufficient to improve performance. Further interventions are necessary.

Mixed-capital enterprise arrangements are also conducive to increased stakeholder involvement. These can be established either by selling a minority or noncontrolling equity stake to private investors (either a strategic equity partner or shareholders brought in by a partial initial public offering) or through private debt markets. Shareholders (through their voting rights and representatives on the board) and bond holders (through debt covenants) can exercise considerable influence. Credit agencies provide financial discipline over managers, who fear a credit downgrading and an increase in capital costs.

Customer-owned enterprises (such as cooperatives and mutuals) are another option. Customers have mandatory representation on boards of directors. Unfortunately, obstacles to collective action can minimize the influence of many small customers, and they can also be susceptible to capture by large customers or special interests. Effective customer governance is more likely in small groups with stable membership and adjacent interests. Cooperatives are more appropriate for smaller, local utilities.

Finally, the most effective way to change the political economy of state-owned electricity utilities is structural reform and the introduction of competition. The potentially competitive elements of the industry (generation and retail) can be separated from the natural monopoly elements of the value chain (the transmission and distribution networks). This can be done piecemeal, first by creating separate business units, which are then transformed into separate companies, with competition whenever possible. Increasing the number of industry players and introducing private sector participation allows for comparisons to be made among the performance of these different entities. Customers can choose their suppliers, and investors and employees of competing firms are incentivized to improve performance. The potential for full retail competition in the power sector in Africa may be limited, but consideration could be given to at least allowing large customers to choose among the incumbent utility and alternative independent power producers or even cross-border imports.

Practical Tools for Improving the Performance of State-Owned Utilities

In addition to governance reforms, practical operational tools have been developed for improving the performance of state-owned utilities. The Commercial Reorientation of the Electricity Sector Toolkit (CREST) is an experiment underway in several localities served by West African electricity providers. Based on good practices from recent reforms in Indian, European, and U.S. power corporations, CREST is a "bottom-up" approach designed to address system losses, low collection rates, and poor customer service. A combination of technical improvements (such as replacing low-tension with high-tension lines and installing highly reliable armored and aerial bunched cables on the low-tension consumer point to reduce theft) and managerial changes (introducing "spot billing" and combining the four transactions of recording, data transfer, bill generation, and distribution) reduces transaction times and generates more regular cash flow (Tallapragada 2008). Early applications of CREST have reportedly produced positive changes in several neighborhoods in Guinea and Nigeria, which are two difficult settings. The application of the toolkit should be closely monitored and evaluated and, if successful, should be replicated elsewhere (Nellis 2008).

Conclusion

Institutional reform is a lengthy process. Victories on this front will be small and slow in coming. Donors may prefer large and quick solutions, but they must recognize that governance reform of state-owned utilities is essential to improving the performance of the African power sector. A key challenge in the sector is funding for new power infrastructure. Improved financial performance of state-owned utilities helps reduce the funding gap by reducing inefficiencies and losses and improving collection rates, revenue, and retained earnings, which can be directed to investments in new capacity or network expansion. Improved performance can also lead to better credit ratings, thereby increasing utilities' access to private debt markets. Improved credit worthiness also means that state-owned utilities can be more reliable counterparties to independent power producer investors, thus once again increasing investment flows into the sector. Improved state-owned utility performance is thus key to meeting the funding challenges outlined in the next chapter.

References

Briceño-Garmendia, Cecilia, and Maria Shkaratan. 2010. "Power Tariffs: Caught between Cost Recovery and Affordability." Working Paper 8, Africa Infrastructure Country Diagnostic, World Bank, Washington, DC.

Eberhard, Anton, Vivien Foster, Cecilia Briceño-Garmendia, Fatimata Ouedraogo, Daniel Camos, and Maria Shkaratan. 2008. "Underpowered: The State of the Power Sector in Sub-Saharan Africa." Background Paper 6, Africa Infrastructure Sector Diagnostic, World Bank, Washington, DC.

Foster, Vivien, and Cecilia Briceño-Garmendia, eds. 2009. *Africa's Infrastructure: A Time for Transformation*. Paris, France, and Washington, DC: Agence Française de Développement and World Bank.

Gomez-Ibanez, J. A. 2007. "Alternatives to Infrastructure Privatization Revisited: Public Enterprise Reform from the 1960s to the 1980s." Policy Research Working Paper 4391, World Bank, Washington, DC.

Molefhi, B. O. C., and L. J. Grobler. 2006. "Demand-Side Management: A Challenge and Opportunity for Botswana Electric Energy Sector." North West University, Potchefstroom, South Africa.

Nellis, John. 2008. "Private Management Contracts in Power Sector in Sub-Saharan Africa." Internal note, Africa Infrastructure Country Diagnostic, World Bank, Washington, DC.

Tallapragada, Prasad V. S. N. 2008. "Commercial Reorientation of the Electricity Sector Toolkit: A Methodology to Improve Infrastructure Service Delivery." Unpublished note, World Bank, Washington, DC.

Vagliasindi, Maria. 2008. "Institutional Infrastructure Indicators: An Application to Reforms, Regulation and Governance in Sub-Saharan Africa." Unpublished paper, AFTSN, World Bank, Washington, DC.

CHAPTER 7

Closing Africa's Power Funding Gap

The cost of addressing Africa's power sector needs is estimated at $40.8 billion a year, equivalent to 6.35 percent of Africa's gross domestic product (GDP). The burden varies greatly by country, from 0.3 percent of GDP in Equatorial Guinea to 35.4 percent in Zimbabwe. Approximately two-thirds of the total spending need is capital investment ($26.7 billion a year); the remainder is operations and maintenance (O&M) expenses ($14.1 billion a year). The model used to calculate these estimates was run under the assumption of expanded regional power trade and takes into account all investments needed for the increase in trade and all cost savings achieved as a result.

In comparison with other sectors, power sector investment needs are very high: They are 4.5 times larger than in the information and communication technology (ICT) sector and approximately double the investment needs in each of the water, sanitation, and transport sectors.

Current spending aimed at addressing power infrastructure needs is higher than previously thought and adds up to an estimated $11.6 billion. Almost equal shares of this amount are spent by three groups of countries: middle-income, resource-rich, and nonfragile low-income countries. Fragile low-income countries spend the remaining small share (5 percent, or approximately $0.83 billion), a reflection of the weakness of their economies. The majority of spending is channeled through public institutions, most notably power sector utilities (state-owned enterprises [SOEs]).

Approximately 80 percent of existing spending is domestically sourced from taxes or user charges. The rest is split among official development assistance (ODA) financing, which provides 6 percent of the total; funding from countries outside the Organisation for Economic Co-operation and Development (OECD), which provides 9 percent of the total, and private sector contributions, which provide 4 percent of the total. Almost 75 percent of domestic spending goes to O&M. Capital spending is financed from four sources: One-half comes from the domestic public sector, approximately one-quarter is received from non-OECD financiers, and the rest is contributed by OECD and the private sector.

Much can be done to reduce the gap between spending needs and current levels of spending. Inefficiencies of various kinds total 1.28 percent of GDP. Reducing inefficiencies is a challenging task, but the financial benefit can be substantial.

Three types of power sector inefficiencies are found. First, there are utility inefficiencies, which include system losses, undercollection of revenue, and overstaffing. These result in a major waste of resources that adds up to $4.40 billion a year. Undercollection, the largest component of utility inefficiencies, amounts to $1.73 billion; system losses account for $1.48 billion, and overstaffing for $1.15 billion.

The second type of sector inefficiency is underpricing of power. By setting tariffs below the levels needed to cover actual costs, countries in Sub-Saharan Africa forego revenue of $3.62 billion a year.

The third type of inefficiency is poor budget execution, with only 66 percent of the capital budgets allocated to power actually spent. That leaves an estimated $258 million in public investment that is earmarked for the power sector but diverted elsewhere in the budget.

Tackling all these inefficiencies would make an additional $8.24 billion available, but a funding gap of $20.93 billion would still remain. The situation differs by country; one-third of countries in Sub-Saharan Africa would be able to fund their needs, but the remaining two-thirds would face a funding gap of between 6 and 74 percent of total needs even if all inefficiencies were eliminated.

The countries in the second group will therefore need to pursue ways to raise additional funds. Historical trends do not suggest strong prospects for increasing allocations from the public budget: Even when fiscal surpluses existed, they did not visibly favor infrastructure. External finance for infrastructure has been buoyant in recent years; in particular, funding from OECD has increased. However, the power sector has not benefited from this trend: It has received the least funding compared with transport, water supply and sanitation (WSS), and ICT.

Closing Africa's power infrastructure funding gap inevitably requires reforms to reduce or eliminate inefficiencies. This will help existing resources to go farther and create a more attractive investment climate for external and private finance.

Existing Spending in the Power Sector

Existing spending on infrastructure in Africa is higher than previously thought when the analysis takes into account budget and off-budget spending (including SOEs and extra budgetary funds) and spending financed by external sources including ODA, official sources in non-OECD countries, and private sources.

Africa is spending $11.6 billion a year to address its power infrastructure needs, which is equivalent to 1.8 percent of GDP. This is split between investment (40 percent of the total) and O&M. Although the public sector more or less covers O&M needs, it provides only 51.5 percent of investment financing needs. The rest of investment spending is provided by external and private sector investors.

Of the total investment funds provided by the public sector for infrastructure, power amounts to one-quarter, transport nearly one-half, and the remaining one-quarter is divided more or less equally between the WSS and ICT sectors (table 7.1). The power sector receives nearly half of the infrastructure funding provided by non-OECD financiers but does

Table 7.1   Sectoral Composition of Investment, by Financing Source
percent

	Power	Transport	WSS	ICT	Irrigation
Domestic public sector	24	47	13	13	3
External and private sector	14	25	23	37	0
Including					
ODA	19	48	33	0	0
Non-OECD	47	46	7	0	0
Private	5	11	23	61	0

Source: Briceño-Garmendia, Smits, and Foster (2008) for public spending, PPIAF (2008) for private flows, and Foster and others (2008) for non-OECD financiers.
Note: All rows total 100 percent. ICT = information and communication technology; ODA = official development assistance; OECD = Organisation for Economic Co-operation and Development; WSS = water supply and sanitation.

less well in ODA and PPI funding. Telecommunications receives the majority of private infrastructure funding.

Funding patterns vary considerably across countries, which is explained in part by the economic and political status of each country. We can group countries into four broad categories to make sense of these variations: middle-income countries, resource-rich countries, fragile low-income countries, and nonfragile low-income countries (box 7.1).

Middle-income and resource-rich countries spend 1.3 percent and 1.8 percent of GDP on power, respectively. Low-income countries spend substantially more: 2.2 percent of GDP in the nonfragile states and 2.9 percent of GDP in fragile states (table 7.2). The composition of spending also varies substantially across country groups. Middle-income countries allocate three-quarters of power spending to O&M; this is the case primarily

---

Box 7.1

Introducing a Country Typology

Middle-income countries have GDP per capita in excess of $745 but less than $9,206. Examples include Cape Verde, Lesotho, and South Africa (World Bank 2007).

Resource-rich countries are low-income countries whose behaviors are strongly affected by their endowment of natural resources (Collier and O'Connell 2006; IMF 2007). Resource-rich countries typically depend on exports of minerals, petroleum, or both. A country is classified as resource rich if primary commodity rents exceed 10 percent of GDP. (South Africa is not classified as resource-rich, using this criterion). Examples include Cameroon, Nigeria, and Zambia.

Fragile states are low-income countries that face particularly severe development challenges, such as weak governance, limited administrative capacity, violence, or a legacy of conflict. In defining policies and approaches toward fragile states, different organizations have used differing criteria and terms. Countries that score less than 3.2 on the World Bank's Country Policy and Institutional Performance Assessment belong to this group. Fourteen countries of Sub-Saharan Africa are in this category. Examples include Côte d'Ivoire, the Democratic Republic of Congo, and Sudan (World Bank 2005).

Other low-income countries constitute a residual category of countries that have GDP per capita below $745 and are neither resource rich nor fragile states. Examples include Benin, Ethiopia, Senegal, and Uganda.

(continued next page)

Box 7.1 *(continued)*

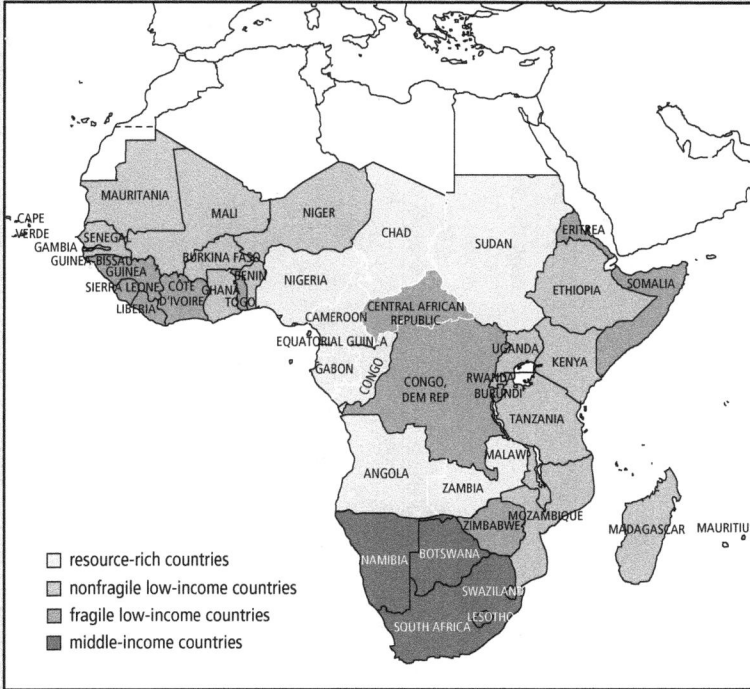

Source: Briceño-Garmendia, Smits, and Foster 2008.

because the largest, South Africa, has been delaying investment in new capacity. Fragile low-income countries spend 70 percent on O&M, and nonfragile low-income countries allocate 60 percent of the power budget to O&M. By contrast, resource-rich countries spend only 40 percent on O&M and allocate the rest to investment.

The variation of power sector spending across countries ranges from less than 0.1 percent of GDP in the Democratic Republic of Congo to almost 6 percent of GDP in Cape Verde (figure 7.1a). Countries with low levels of capital spending include Lesotho (0.10 percent of GDP), South Africa (0.27 percent of GDP), Madagascar (0.36 percent of GDP), and Malawi (0.56 percent of GDP). All these countries require additional investment in new generation capacity or power transmission (Rosnes and Vennemo 2008). At the other end of the scale are countries with high

Table 7.2 Power Sector Spending in Sub-Saharan Africa, Annualized Flows

	Percentage of GDP							$ million						
	O&M	Capital expenditure						O&M	Capital expenditure					
Country type	Public sector	Public sector	ODA	Non-OECD financiers	PPI	Total capital expenditures	Total	Public sector	Public sector	ODA	Non-OECD financiers	PPI	Total capital expenditures	Total
Middle income	0.98	0.28	0.01	0.00	0.00	0.30	1.28	2,656	772	33	1	5	811	3,467
Resource rich	0.72	0.56	0.03	0.33	0.13	1.05	1.77	1,602	1,243	75	736	278	2,333	3,935
Nonfragile low income	1.78	0.39	0.50	0.12	0.15	1.15	2.94	1,970	432	549	129	165	1,274	3,243
Fragile low income	1.49	0.00	0.10	0.55	0.03	0.68	2.16	571	0	37	210	12	260	830
Africa	1.09	0.37	0.11	0.17	0.07	0.72	1.81	7,011	2,363	694	1,076	460	4,594	11,605

Source: Briceño-Garmendia, Smits, and Foster (2008) for public spending; PPIAF (2008) for private flows; Foster and others (2008) for non-OECD financiers.

Note: Aggregate public sector expenditure covers general government and state-owned enterprise expenditure on infrastructure. Figures are extrapolations based on the 24-country sample covered in AICD Phase 1. Totals may not add exactly because of rounding errors. GDP = gross domestic product; ODA = official development assistance; OECD = Organisation for Economic Co-operation and Development; O&M = operation and maintenance; PPI = private participation in infrastructure.

levels of capital expenditure. This group includes Uganda (3.1 percent of GDP) and Ghana (1.4 percent of GDP).

The funding received from different sources also varies substantially across countries (figure 7.1b). Although public funding is the dominant source in 83 percent of countries, ODA plays a substantial role in many low-income countries. A handful of countries enjoy a significant contribution from the private sector. Non-OECD finance contributes a relatively small amount to the power sector in most countries, with the exception of Ghana and Niger, where it exceeds 20 percent of the total.

Figure 7.1   Power Spending from All Sources as a Percentage of GDP

a. By functional category

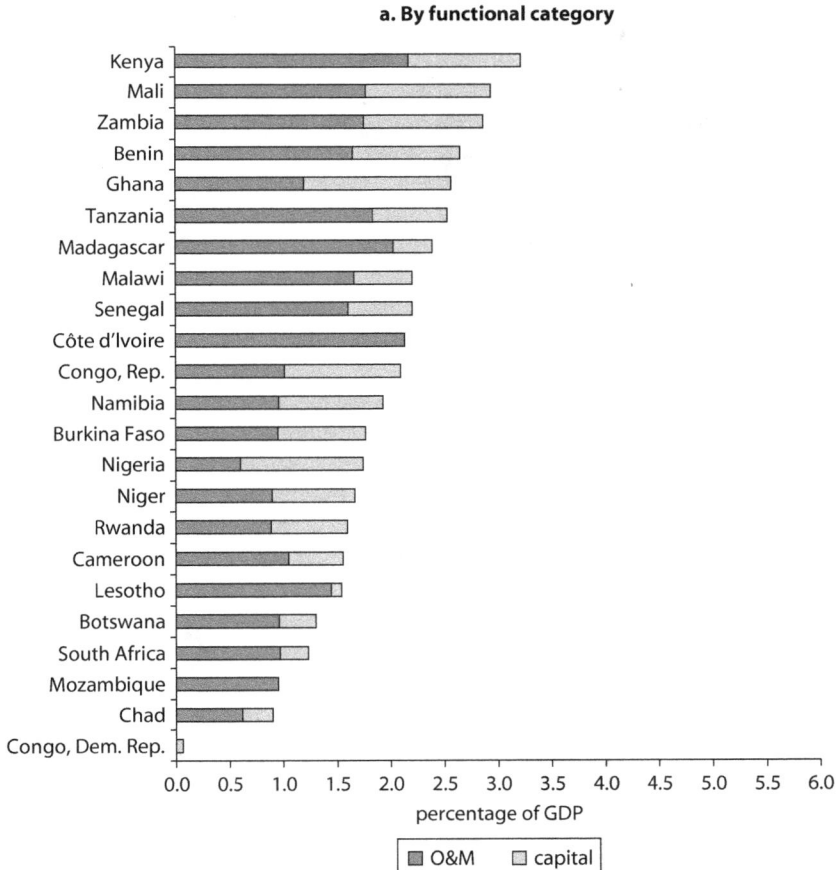

percentage of GDP

■ O&M   □ capital

(continued next page)

Figure 7.1   *(continued)*

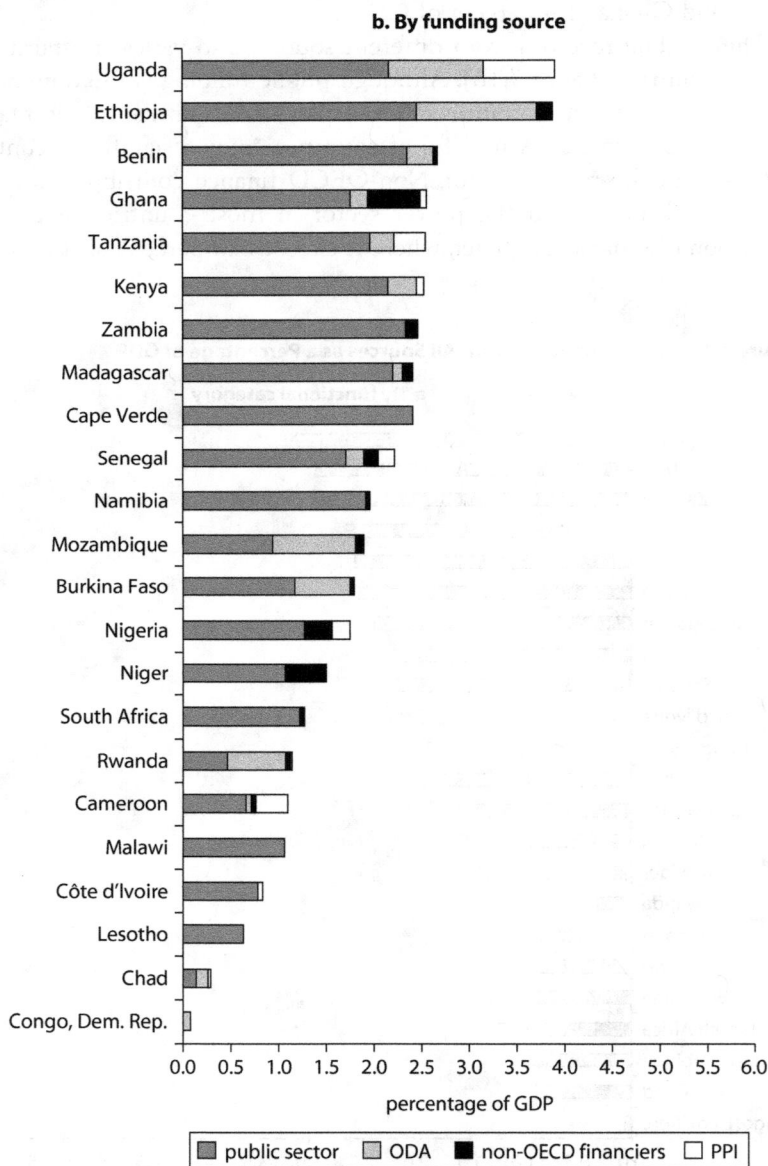

b. By funding source

percentage of GDP

■ public sector   ▦ ODA   ■ non-OECD financiers   ☐ PPI

Source: Briceño-Garmendia, Smits, and Foster (2008) for public spending, PPIAF (2008) for private flows, Foster and others (2008) for non-OECD financiers.
Note: Based on annualized averages for 2001–06. Averages weighted by country GDP. GDP = gross domestic product; ODA = official development assistance; OECD = Organisation for Economic Co-operation and Development; O&M = operations and maintenance; PPI = private participation in infrastructure.

In the middle-income countries, domestic public sector resources (including tax revenue and user charges raised by state entities) account for 99 percent of power sector spending. Across the other country categories, domestic public sector resources invariably contribute at least two-thirds of total spending. In the middle- and low-income countries, domestic public spending is focused on O&M, which accounts for more than three-quarters of the total. In the resource-rich states, domestic public spending in the power sector is more balanced, with only 57 percent of the total spent on O&M.

In the aggregate, external finance contributes roughly one-half of Africa's total capital spending on the power sector. External sources include ODA, official finance from non-OECD countries (such as China, India, and the Arab funds), and PPI. External finance is primarily for investment—broadly defined to include asset rehabilitation and construction—and does not provide for O&M. One-half of external finance for Africa's power sector comes from non-OECD financiers, approximately one-third from PPI, and the rest, roughly 20 percent, from ODA (table 7.1).

External financing favors resource-rich countries: They obtain approximately 50 percent of total external funds. The second largest recipient of external financing is the group of nonfragile low-income countries, which receive one-third of the total. ODA is directed primarily (80 percent) to nonfragile low-income states. Two-thirds of financing from each of the other two sources—non-OECD financiers and PPI—benefits resource-rich countries (figure 7.2).

How Much More Can Be Done within the Existing Resource Envelope?

Africa is losing an estimated $8.24 billion per year to various inefficiencies in its power sector. In this context, four distinct opportunities can be identified for efficiency gains. The lack of cost recovery is the largest source of sector inefficiency: Losses from pricing power below the current costs constitute 44 percent of all inefficiencies. Essential interventions include improving utility operations, capitalizing on the benefits of regional trade, and bringing tariffs to the level of the long-run marginal costs of power. Undercollection of bills adds up to 22 percent of total sector inefficiency, and the utilities should tackle this issue. System losses constitute 18 percent of the inefficiencies and need to be addressed. Overstaffing in the power utilities contributes 14 percent of total inefficiencies.

Figure 7.2   Sources of Financing for Power Sector Capital Investment

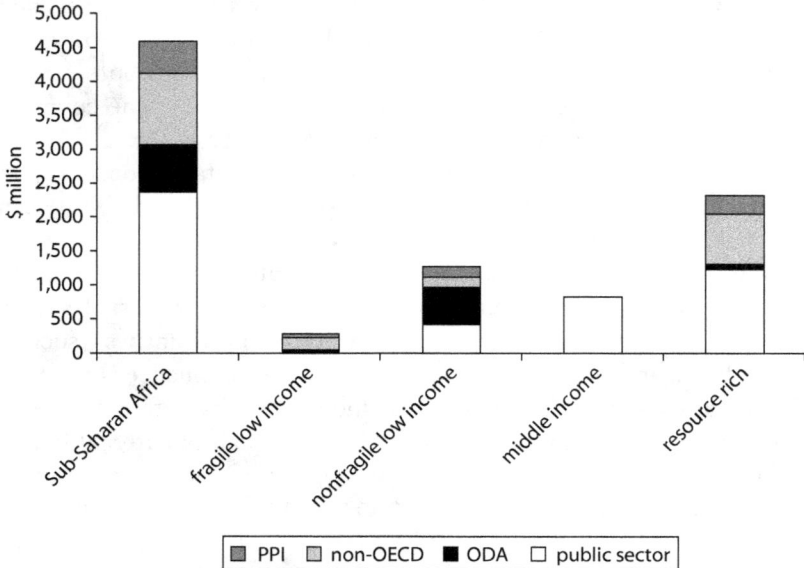

Source: Briceño-Garmendia, Smits, and Foster (2008) for public spending; PPIAF (2008) for private flows; Foster and others (2008) for non-OECD financiers.
Note: ODA = official development assistance; OECD = Organisation for Economic Co-operation and Development; PPI = private participation in infrastructure.

Increasing Cost Recovery

By setting tariffs below the levels needed to recover actual costs, Sub-Saharan countries forego revenue of $3.62 billion a year.[1] However, low cost recovery is a function of both low tariffs and high costs. Despite comparatively high power prices, most Sub-Saharan Africa countries are recovering only their average operating costs and are far from being able to recover total costs with tariffs. Although a few countries—Burkina Faso, Cape Verde, Chad, Côte d'Ivoire, Kenya, Namibia, and Uganda—achieved cost recovery, they are exceptions. Also, in some cases (Burkina Faso, Cape Verde, and Chad), cost recovery has been achieved by elevating tariffs above extremely high costs. Power tariffs in Sub-Saharan Africa are high compared with other regions. The average power tariff of $0.12 per kilowatt-hour (kWh) is twice the level in other developing regions, such as South Asia. The high costs of power can, to a large extent, be explained by lack of economies of scale, underdeveloped regional energy resources, high oil prices, drought, and political instability—factors mostly beyond the influence of the energy sector or a utility. However, other causes could be resolved at the sector or utility level. One example is subsidized residential tariffs, especially in the countries with a high share of residential consumption. Another is inefficient residential tariff structures that decrease with increased consumption and create cross-subsidies from the lower-income households to the more affluent ones, which curtails usage by poorer households and promotes overconsumption of power by more affluent households.

When tariffs charged to residential customers are below costs (figure 7.3), motivating and achieving increases is usually socially and politically sensitive and takes time to accomplish. In addition, many countries in Sub-Saharan Africa are pricing power to highly energy-intensive industries at greatly subsidized rates. These arrangements were initially justified as ways of locking in baseload demand to support the development of very large-scale power projects that went beyond the immediate demands of the country, but they have become increasingly questionable as competing demands have grown to absorb this capacity. Salient examples include the aluminum-smelting industry in Cameroon, Ghana, and South Africa and the mining industry in Zambia.

As figure 7.3 demonstrates, total costs of power supply are above the average tariffs for all customer groups, including residential and industrial tariffs. On average, total costs exceed residential tariffs by 23 percent and industrial tariffs by 36 percent.

Figure 7.3   Power Prices and Costs, Sub-Saharan Africa Average

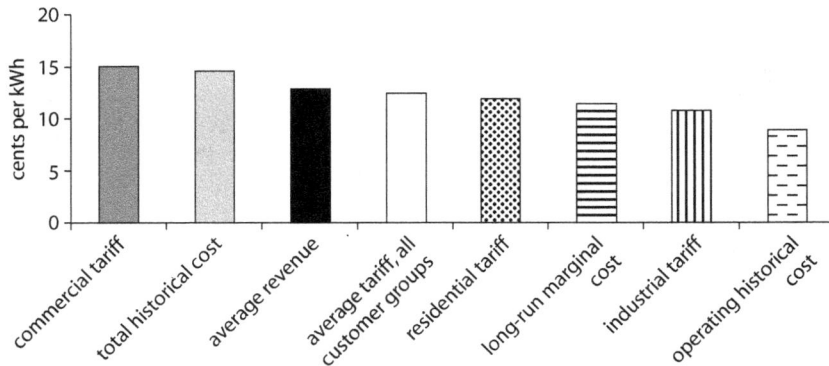

Source: Briceño-Garmendia and Shkaratan 2010.
Note: kWh = kilowatt-hour.

Although tariffs in most countries fall below total costs, they recover operational costs, with only a few exceptions. The exceptions include Cameroon, Malawi, Niger, Tanzania, and Zambia. On the aggregate continental level, tariffs are 40 percent above the operational cost level. Differences are seen among customer groups in this respect: Commercial, residential, and industrial tariffs exceed operational costs by 67 percent, 43 percent, and 21 percent, respectively.

On Budget Spending: Raising Capital Budget Execution

As mentioned previously, most public spending in the power sector SOEs in Sub-Saharan Africa is off-budget, while the on-budget spending constitutes only a small portion of it. The public sector in Sub-Saharan Africa allocates 0.13 percent of GDP, or $827 million, to support the power sector (table 7.3). For a typical African country, this effort translates to about $29 million a year, which is very small in relation to overall power sector needs. To put this figure in perspective, the power sector needs in Sub-Saharan African countries range from $2 million to $13.5 billion per year, and budgetary support of the sector varies from zero to $444 million. Although 99.6 percent of power sector public spending in the middle-income countries is off budget, for resource-rich countries, off-budget spending is a much smaller part of the total, equal to 71.2 percent of all public resources dedicated to power.

Despite the limited allocation of public budgetary spending to the power sector, it is still important to mention one additional source of inefficiency: poor budget execution. Central governments face significant problems in executing their infrastructure budgets. On average, African

Table 7.3   Annual Budgetary Flows to Power Sector

Country type	Percentage of GDP		$ billion	
	Power	Total	Power	Total
Resource rich	0.37	1.60	0.815	3.55
Middle income	0.01	1.46	0.015	3.96
Nonfragile low income	0.13	1.52	0.145	1.67
Fragile low income	0.0	0.71	0.0	0.27
Africa	0.13	1.48	0.827	9.50

Source: Briceño-Garmendia, Smits, and Foster 2008.
Note: Based on annualized averages for 2001–06. Averages weighted by country GDP. Figures are extrapolations based on the 24-country sample covered in AICD Phase 1. Totals may not add exactly because of rounding errors. GDP = gross domestic product.

Table 7.4   Average Budget Variation Ratios for Capital Spending

Country type	Overall infrastructure	Power
Middle income	78	—
Resource rich	65	60
Nonfragile low income	76	75
Fragile low income	—	—
Sub-Saharan Africa	75	66

Source: Adapted from Briceño-Garmendia, Smits, and Foster 2008.
Note: Based on annualized averages for 2001–06. — = data not available.

countries are unable to spend as much as one-third of their capital budgets for power (table 7.4). The poor timing of project appraisals and late releases of budgeted funds due to procurement problems often prevent the use of resources within the budget cycle. Delays affecting in-year fund releases are also associated with poor project preparation, which leads to changes in the terms agreed upon with contractors in the original contract (such as deadlines, technical specifications, budgets, and costs). In other cases, cash is reallocated to nondiscretionary spending driven by political or social pressures.

Unlike in other infrastructure sectors, the power sector's losses from nonexecution of budgets are small as a percentage of spending. However, the absolute amount is large, and it is important to tackle this inefficiency. If the bottlenecks in power sector capital execution could be resolved, countries would increase their spending on power by $258 million a year, or 2.2 percent of total current spending, without any increase in current budget allocations. Resolution of these planning, budgeting, and procurement challenges should be included in the region's reform agenda.

Even if budgets are fully spent, concerns are found as to whether funds reach their final destinations. Public expenditure tracking surveys have attempted to trace the share of each budget dollar that results in productive frontline expenditures. Most of the existing case studies concern social sectors as opposed to power, but they illustrate leakages of public capital spending that can be as high as 92 percent (see Pritchett 1996; Rajkumar and Swaroop 2002; Reinikka and Svensson 2002, 2004; Warlters and Auriol 2005).

Improving Utility Performance

Utility inefficiencies are high and constitute on average 0.68 percent of GDP in Sub-Saharan African countries. In some countries, inefficiencies

amount to almost 5 percent of GDP. Looking at different sources of utility inefficiency, one can see that the largest component is undercollection of electricity bills (0.40 percent of GDP), followed by system losses
(0.34 percent of GDP) and overstaffing at the SOEs (0.26 percent of
GDP). These numbers are monetary equivalents of physical measures of
inefficiencies, such as system losses that average 23 percent compared
with a global norm of 10 percent. Collection rates average 88.4 percent
compared with the best practice standard of 100 percent, and customer-
to-employee ratios in Sub-Saharan Africa average 184, substantially
below the same indicator in other developing regions.

In countries with above-average utility inefficiencies, growth in power
access is slow and suppressed demand high compared with the rest of the
countries. If revenue cannot cover the necessary expenses because of
undercollection or system losses, or the salary bill is excessively high,
government resources are used to subsidize the utility. When subsidies
cannot cover the net loss, the utilities are forced to skimp on maintenance, and performance deteriorates even further.

Savings from Efficiency-Oriented Reforms

In total, $8.2 billion could be captured through efficiency improvements,
cost recovery, and more effective budget execution. The largest potential
gains come from improved operational efficiencies that amount to $4.4
billion a year, most of which would come from achieving a 100 percent
collection rate ($1.7 billion). A further $1.5 billion a year could be secured
by reducing system losses to the internationally recognized norm. Dealing
with overstaffing would liberate another $1.2 billion (table 7.5). Reaching
cost recovery through cost reduction and tariff adjustment, as described

Table 7.5   Potential Gains from Higher Operational Efficiency
$ million annually

	Middle income	Resource rich	Nonfragile low income	Fragile low income	Total Sub-Saharan Africa
All operational inefficiencies	1,745	1,838	980	1,738	4,355
System losses	22	948	470	498	1,476
Undercollection	96	480	339	1,141	1,728
Overstaffing	1,627	410	172	99	1,152

Source: Briceño-Garmendia, Smits, and Foster 2008.

earlier, would yield $3.6 billion. Finally, achieving full capital execution would add yet another 0.2 billion a year.

Ten countries have potential efficiency savings of more than 2 percentage points of GDP, from as much as 4.5 percent of GDP in the case of Côte d'Ivoire to 2.38 percent of GDP in Ghana. An additional eight countries can potentially save 1–2 percent of their GDP by eliminating inefficiencies (figure 7.4). In 56 percent of the countries, the largest

Figure 7.4   Potential Efficiency Gains from Different Sources

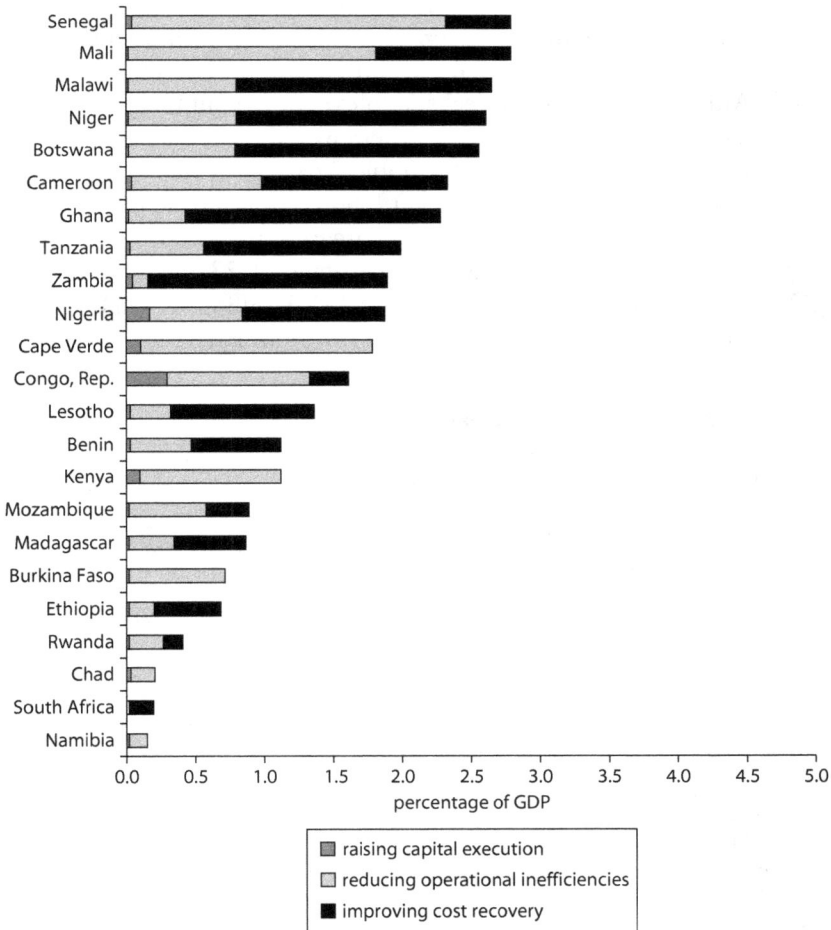

percentage of GDP

- ▨ raising capital execution
- ☐ reducing operational inefficiencies
- ■ improving cost recovery

Source: Briceño-Garmendia, Smits, and Foster 2008.
Note: Based on annualized averages for 2001–06. Averages weighted by country GDP. GDP = gross domestic product.

source of inefficiencies is the lack of cost recovery. Operational deficiencies are the main source of inefficiency in 44 percent of the countries.

Annual Funding Gap

Existing spending and potential efficiency gains can be subtracted from
estimated spending needs to gauge the extent of the shortfall in funding.
However, even if all of the inefficiencies described previously could be
tackled, they would cover only 20 percent of the funding gap for the
power sector in Sub-Saharan Africa, 38 percent in resource-rich and
low-income fragile states, 18 percent in nonfragile low-income countries,
and just 17 percent in middle-income countries. About three-quarters of
this funding gap relates to capital investment, and the remainder is O&M
needs. Although it may be unrealistic to expect that all these inefficiencies could be captured, even halving them would make a contribution to
financing the African power sector (table 7.6).

Seventeen countries face significant funding gaps for the power sector
(figure 7.5). By far the most salient cases are Ethiopia and the Democratic
Republic of Congo, which have annual gaps of 23 percent of GDP
($2.8 billion annually) and 18 percent ($1.3 billion a year), respectively. Mozambique, Senegal, and Madagascar all have funding gaps of

Table 7.6   Annual Power Funding Gap

	Middle income	Resource rich	Nonfragile low income	Fragile low income	Total Sub-Saharan Africa	Cross-country gain from reallocation
Infrastructure spending needs	(14,191)	(11,770)	(9,704)	(5,201)	(40,797)	n.a.
Spending directed to needs	3,470	3,959	3,241	830	11,633	n.a.
Gain from eliminating inefficiencies	2,431	4,440	1,758	1,924	8,237	n.a.
Capital execution	2	294	20	0	258	n.a.
Operational inefficiencies:	1,745	1,838	980	1,738	4,355	n.a.
Cost recovery	684	2,309	757	186	3,624	n.a.
Funding gap	(8,289)	(3,370)	(4,705)	(2,447)	(20,927)	n.a.
Potential for reallocation	0	0	0	0	0	773

Source: Briceño-Garmendia, Smits, and Foster 2008.
Note: n.a. = Not applicable.

Figure 7.5   Power Infrastructure Funding Gap

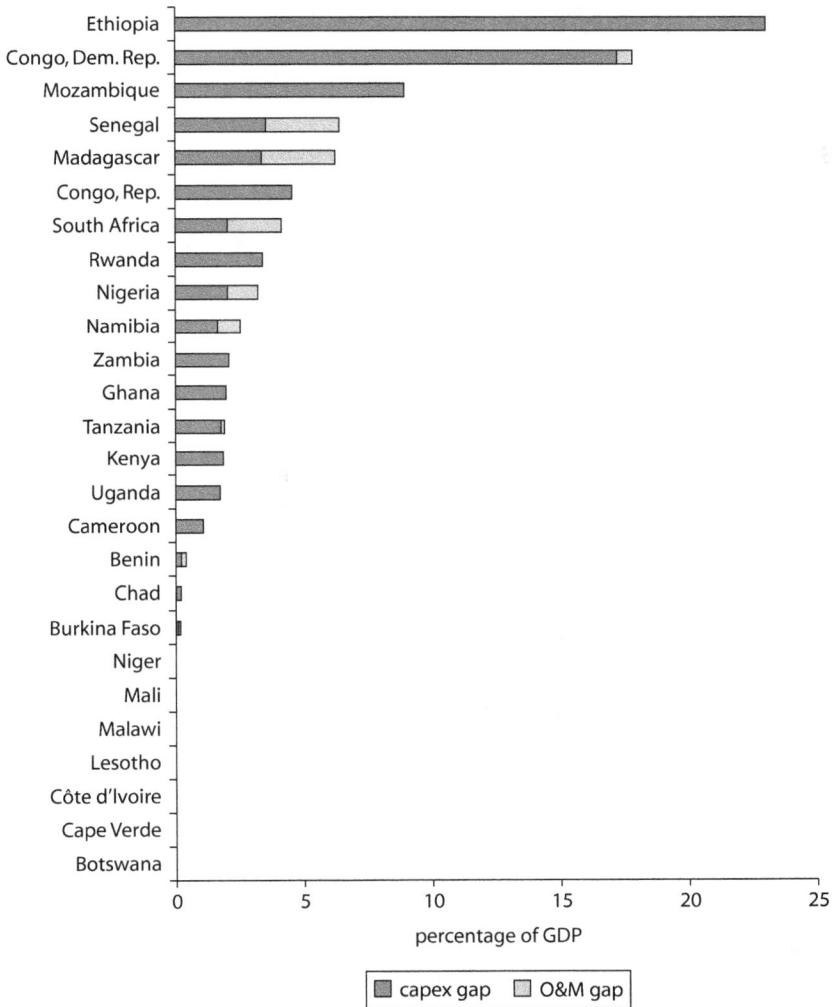

Source: Briceño-Garmendia, Smits, and Foster 2008.
Note: Based on annualized averages for 2001–06. Averages weighted by country GDP. capex = capital expenditures; GDP = gross domestic product; O&M = operations and management.

5–10 percent of GDP. The Democratic Republic of Congo, South Africa, Rwanda, Nigeria, Namibia, Zambia, Ghana, Tanzania, Kenya, Uganda, and Cameroon have funding gaps of 1–5 percent of GDP.

After inefficiencies are eliminated, the power sector's annual funding gap totals $20.9 billion. Covering it would require raising additional funds, taking more time to attain investment and coverage targets, or using lower-cost technologies. The remainder of this chapter evaluates the potential for raising additional finance and explores policy adjustments to reduce the price tag and the burden of the funding gap.

How Much Additional Finance Can Be Raised?

Only limited financing sources are available, and the 2008 global financial crisis has affected all of them adversely. Domestic public finance is the largest source of funding today, but it presents little scope for an increase, except possibly in countries enjoying natural resource windfalls. The future of ODA and non-OECD financing is unclear in the postcrisis situation. Although private participation in the power sector in Africa has increased over the past two decades, it remains at modest levels, and investors are more cautious after the 2008 financial crisis. The question is whether private participation might increase in the future, assuming capacity expansion, an improved institutional environment, and reduced barriers to entry. Local capital markets have so far contributed little to infrastructure finance outside South Africa, and to a smaller extent in Kenya, but they could eventually become more important in some of the region's larger economies. Moreover, both of these sources of funding are of limited relevance to the power sector in fragile low-income states, which is where public resources are least available.

Little Scope for Raising More Domestic Finance

To what extent are countries willing to allocate additional fiscal resources to infrastructure? In the run-up to the current financial crisis, the fiscal situation in Sub-Saharan Africa was favorable. Rapid economic growth, averaging 4 percent a year from 2001 to 2005, translated into increased domestic fiscal revenue of about 3 percent of GDP on average. In resource-rich countries, burgeoning resource royalties added 7.7 percent of GDP to the public budget. In low-income countries, substantial debt relief increased external grants by almost 2 percent of GDP.

To what extent were the additional resources available during the recent growth surge allocated to infrastructure? The answer is: surprisingly little (table 7.7). The most extreme case is that of the resource-rich countries, particularly Nigeria. Huge debt repayments more than fully absorbed the fiscal windfalls in these countries. As a result, budgetary spending actually contracted by 3.7 percent of GDP. Infrastructure

Table 7.7   Net Change in Central Government Budgets, by Economic Use, 1995–2004
percentage of GDP

Use	Sub-Saharan Africa	Middle income	Resource rich	Fragile low income	Nonfragile low income
Net expenditure budget	1.89	4.08	(3.73)	1.69	3.85
Current infrastructure spending as a share of expenditures	0.00	0.02	0.03	0.00	0.09
Capital infrastructure spending as a share of expenditures	(0.14)	0.04	(1.46)	0.54	0.22

Source: Adapted from Briceño-Garmendia, Smits, and Foster 2008.
Note: Based on annualized averages for 2001–06. Averages weighted by country GDP. Totals are extrapolations based on the 24-country sample as covered in AICD Phase 1. GDP = gross domestic product.

investment, which bore much of the decrease in spending, fell by almost 1.5 percent of GDP. In middle-income countries, budgetary spending increased by almost 4.1 percent of GDP, but the effect on infrastructure spending was almost negligible, and the additional resources went primarily to current social sector spending. Only in the low-income countries did the overall increases in budgetary expenditure have some effect on infrastructure spending. Even there, however, the effect was fairly modest and confined to capital spending. The nonfragile low-income countries have allocated 30 percent of the budgetary increase to infrastructure investments. The fragile states, despite seeing their overall budgetary expenditures increase by about 3.9 percent of GDP, have allocated only 6 percent of the increase to infrastructure.

Compared with other developing regions, Sub-Saharan Africa's public financing capabilities are characterized by weak tax revenue collection. Domestic revenue generation of approximately 23 percent of GDP trails averages for other developing countries and is lowest for low-income countries (less than 15 percent of GDP a year). Despite the high growth rates in the last decade, domestically raised revenue grew by less than 1.2 percent of GDP. This finding suggests that raising domestic revenue above current levels would require undertaking challenging institutional reforms to increase the effectiveness of revenue collection and broaden the tax base. Without such reforms, domestic revenue generation will remain weak.

The borrowing capacity from domestic and external sources is also limited. Domestic borrowing is often very expensive, with interest rates that far exceed those on concessional external loans. Particularly for the poorest countries, the scarcity of private domestic savings means that public domestic borrowing tends to precipitate sharp increases in interest rates that build up a vicious circle. For many Sub-Saharan countries, the ratios of debt service to GDP are more than 6 percent.

The 2008 global financial crisis can be expected to reduce fiscal receipts because of lower revenue from taxes, royalties, and user charges. Africa is not exempt from its impact. Growth projections for the coming years have been revised downward from 5.1 percent to 3.5 percent, which will reduce tax revenue and likely depress the demand and willingness to pay for infrastructure services. Commodity prices have fallen to levels of the early 2000s. The effect on royalty revenue, however, will depend on the saving regime in each country. Various oil producers have been saving royalty revenue in excess of $60 a barrel, so the current downturn will affect savings accounts more than budgets. Overall, this adverse situation created by the global financial crisis will put substantial pressure on public sector budgets. In addition, many African countries are devaluing their currency, reducing the purchasing power of domestic resources.

Based on recent global experience, fiscal adjustment episodes tend to fall disproportionately on public investment—and infrastructure in particular. Experience from earlier crises in East Asia and Latin America indicates that infrastructure spending is vulnerable to budget cutbacks during crisis periods. Based on averages for eight Latin American countries, cuts in infrastructure investment amounted to about 40 percent of the observed fiscal adjustment from the early 1980s to the late 1990s (Calderón and Servén 2004). This reduction was remarkable because public infrastructure investment already represented less than 25 percent of overall public expenditure in Latin American countries. These infrastructure investment cuts were later identified as the underlying problem holding back economic growth in the whole region during the 2000s. Similar patterns were observed in East Asia during the financial crisis of the mid-1990s. For example, Indonesia's total public investment in infrastructure dropped from 6–7 percent of GDP in 1995–97 to 2 percent in 2000. Given recent spending patterns, there is every reason to expect that changes in the overall budget envelope in Africa will affect infrastructure investment in a similar pro-cyclical manner.

Official Development Assistance—Sustaining the Scale-Up
For most of the 1990s and early 2000s, ODA financial flows to power infrastructure in Sub-Saharan Africa remained steady at a meager $492

million a year. The launch of the Commission for Africa Report in 2004 was followed by the Group of Eight Gleneagles Summit in July 2005, where the Infrastructure Consortium for Africa was created to focus on scaling up donor finance to meet Africa's infrastructure needs. Donors have so far lived up to their promises, and ODA commitments to African power infrastructure increased by more than 26 percent, from $642 million in 2004 to $810 million in 2006. Most of this ODA comes from multilateral donors—the African Development Bank, European Community, and International Development Association (IDA)—and France and Japan make significant contributions among the bilaterals. A significant lag occurs between ODA commitments and their disbursement, which suggests that disbursements should continue to increase in the coming years. However, this happens less in the power sector than in other infrastructure sectors. In 2006, the just-reported commitments in power were only 18 percent higher than the estimated ODA disbursements of $694 million (see table 7.1). This gap reflects delays typically associated with project implementation. Because ODA is channeled through the government budget, the execution of funds faces some of the same problems affecting domestically financed public investment, including procurement delays and low country capacity to execute funds. Divergences between donor and country financial systems, as well as unpredictability in the release of funds, may further impede the disbursement of donor resources. Bearing all this in mind, if all commitments up to 2007 are fully honored, ODA disbursements could be expected to rise significantly (IMF 2009; World Economic Outlook 2008).

ODA was set to increase further before the crisis, but prospects no longer look so good. The three multilateral agencies—the African Development Bank, the European Commission, and the World Bank—secured record replenishments for their concessional funding windows for the three to four years beginning in 2008. In principle, funding allocations to African infrastructure totaling $5.2 billion a year could come from the multilateral agencies alone in the near future, and power will likely continue to attract a substantial share of that overall envelope. In practice, however, the crisis may divert multilateral resources away from infrastructure projects and toward emergency fiscal support. Bilateral support, based on annual budget determinations, may be more sensitive to the fiscal squeeze in OECD countries, and some decline can be anticipated. Historical trends suggest that ODA has tended to be pro-cyclical rather than countercyclical (IMF 2009; ODI 2009; UBS Investment Research 2008; World Economic Outlook 2008; and references cited therein).

Non-OECD Financiers—Will Growth Continue?

Non-OECD countries financed about $1.1 billion of the African power sector annually during 2001–06 (see table 7.1). This is substantially more than the $0.7 billion provided by ODA over the same period; moreover, the focus of the finance is very different. Non-OECD financiers have been active primarily in oil-exporting countries (Angola, Nigeria, and Sudan). Non-OECD finance for the African power sector has predominantly taken the form of Chinese funding, followed by Indian and then Arab support.

About one-third of Chinese infrastructure financing in Africa has been directed to the power sector, amounting to $5.3 billion in cumulative commitments by 2007. Most of this has been focused on the construction of large hydropower schemes. By the end of 2007, China was providing $3.3 billion for the construction of 10 major hydropower projects totaling 6,000 megawatts (MW). Some of the projects will more than double the generating capacity of the countries where they are located. Outside hydropower, China has invested in building thermal plants, with the most significant projects in Sudan and Nigeria. Main transmission projects are in Tanzania and Luanda (Angola).

Non-OECD finance raises concerns about sustainability. The non-OECD financiers from China, India, and the Arab funds follow sectors, countries, and circumstances aligned with their national business interests. They offer realistic financing options for power and transport and for postconflict countries with natural resources. However, nongovernmental organizations are voicing concerns about the associated social and environmental standards. Non-OECD financiers also provide investment finance without associated support on the operational, institutional, and policy sides, which raises questions about the new assets' sustainability.

How the current economic downturn will affect non-OECD finance is difficult to predict because of the relatively recent nature of these capital inflows. As they originate in fiscal and royalty resources in their countries of origin, they will likely suffer from budgetary cutbacks. The downturn in global commodity prices may also affect the motivation for some of the Chinese infrastructure finance linked to natural resource development.

Private Investors—Over the Hill

Private investment commitments in the Sub-Saharan power sector surged from $40 million in 1990 to $77 million in 1995, then to $451 million in 2000 and $1.2 billion in 2008. It is important to note that these and all values reported here exclude royalty payments to governments for power infrastructure, which—although valuable from a fiscal perspective—do not contribute to the creation of new power assets. When project implementation cycles are taken into account, this translates to average annual disbursements in recent years of $460 million, or 0.07 percent of GDP (see table 7.1). These disbursements are very similar in magnitude to those received from non-OECD financiers, although their composition differs.

Private capital flows to the African power sector have been volatile over time (figure 7.6a), with occasional spikes driven by the closure of a handful of large deals. Excluding this handful of megaprojects, the typical average annual capital flow to African power sector since 2000 has averaged no more than $450 million.

About 80 percent of private finance for African power has gone to greenfield projects with some $7.7 billion of cumulative commitments, a further 17 percent to concessions that amount to cumulative commitments of $1.6 billion, and the remaining 1 percent to divestitures that total $124 million (figure 7.6b).

Private capital flows, in particular, are likely to be affected by the 2008 global financial crisis. In the aftermath of the Asian financial crisis, private participation in developing countries fell by about one-half over a period of five years following the peak of this participation in 1997. Existing transactions are also coming under stress as they encounter difficulties refinancing short- and medium-term debt.

Local Capital Markets—A Possibility in the Medium Term

The outstanding stock of power infrastructure issues in the local capital markets in Africa is $9.6 billion. This is very little compared with annual power sector financing needs ($40.1 billion) and the funding gap ($22.3 billion). Furthermore, this is barely 13 percent of the total outstanding stock of infrastructure issues. In the power sector, the sources of financing are divided almost equally among corporate bond issues (38 percent of total), equity issues (34 percent of total), and bank loans (28 percent of total). Other than in South Africa, corporate bonds are almost nonexistent. Approximately half of local financing of the power sector comes from loans received from the banks, and the other half is covered by utility-issued securities. In South Africa, the picture is very different: Approximately half of financing is a result of corporate bond issuance, almost one-third comes from issuing securities, and only 18 percent is bank lending.

Figure 7.6   Overview of Private Investment to African Power Infrastructure

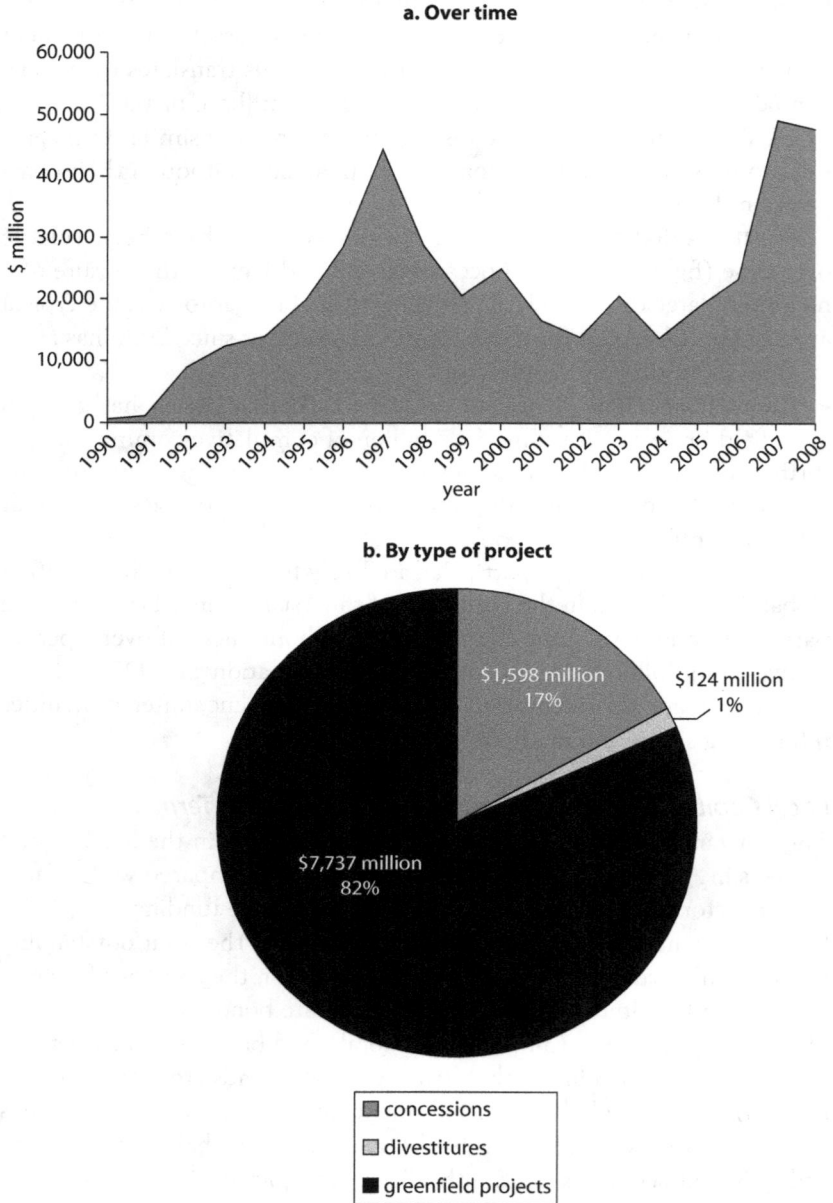

a. Over time

b. By type of project

$1,598 million
17%

$124 million
1%

$7,737 million
82%

- concessions
- divestitures
- greenfield projects

Source: PPIAF 2008.

Although half of the total value of corporate bonds in infrastructure is accounted for in power utilities, only one-quarter of total bank loans to infrastructure goes to the power sector, and only 6 percent of the total value of equity issues is attributed to the power sector (table 7.8).

By comparing countries of different types, one can see that most local capital market financing outside South Africa goes to nonfragile low-income countries (55 percent of total value), and another large part ends up in resource-rich countries (39 percent of total value). Almost the entire value of equities (99 percent of total) is issued in nonfragile low-income countries, and a similar distribution can be observed for corporate bonds, with 88 percent of their value associated with issues in nonfragile low-income countries, although the total value of corporate bonds issued outside South Africa is negligible at $59 million. Most bank loans (68 percent of total) benefit resource-rich countries (table 7.9).

Bank Lending

As of the end of 2006, the amount of commercial bank lending to infrastructure in Africa totaled $11.3 billion. More than $2.7 billion of this total was related to power and water utilities, but distribution between these two sectors was unclear (table 7.8).

As well as being limited in size, bank lending to infrastructure tends to be short in tenure for all but the most select bank clients, which reflects the predominantly short-term nature of banks' deposits and other liabilities. Financial sector officials in Ghana, Lesotho, Namibia, South Africa, Uganda, and Zambia reported maximum maturity terms of 20 years, the longest such maturities among the focus countries. Eight other countries reported maximum loan maturities of "10 years plus," and maximum maturities in four countries were reported as five or more years. Even where 20-year terms are reportedly available, they may not be affordable for infrastructure purposes. In Ghana and Zambia, for example, average lending rates exceed 20 percent because it is difficult to find infrastructure projects that generate sufficient returns to cover a cost of debt that is greater than 20 percent.

For most Sub-Saharan countries, the capacity of local banking systems is too small and constrained by structural impediments to adequately finance infrastructural development. There may be somewhat more potential in this regard for syndicated lending to infrastructure projects with the participation of local banks, which has been on an overall trend of increase in recent years. The volume of syndicated loans to infrastructure borrowers rose steeply from $0.6 billion in 2000 to $6.3 billion in

Table 7.8 Financial Instruments for Locally Sourced Infrastructure Financing

	\$ million					% of total local capital market financing			
	Bank loans	Government bonds[a]	Corporate bonds	Equity issues	Total	Bank loans	Government bonds[a]	Corporate bonds	Equity issues
Africa excluding South Africa									
All infrastructure	5,007.9	46.8	548.1	7,796.2	13,399.0	37	0.30	4	58
Electricity	1,430.9	0.0	58.9	1,302.9	2,792.7	51	0.00	2	47
South Africa									
All infrastructure	6,274.9	754.3	6,841.3	48,148.7	62,019.2	10	1.00	11	78
Electricity	1,263.8	0.0	3,613.9	1,965.4	6,843.1	18	0.00	53	29
Africa total									
All infrastructure	11,282.7	801.1	7,389.4	55,944.9	75,418.1	15	1.00	10	74
Electricity	2,694.7	0.0	3,672.8	3,268.4	9,635.9	28	0.00	38	34

Source: Adapted from Irving and Manroth 2009.

a. The actual amount of government bonds financing infrastructure may be an underestimate, as a specific financing purpose for these bond issues is generally unavailable. Some of the financing raised via these issues may have been allocated toward infrastructure.

Table 7.9   Outstanding Financing for Power Infrastructure, 2006

	Bank loans ($ million)	Corporate bonds ($ million)	Equity issues ($ million)	Total ($ million)	Share of total stock (%)	Share of total infrastructure stock (%)
South Africa	1,264	3,614	1,965	6,843	70	11
Middle income (excluding South Africa)	103	—	—	103	1	19
Resource rich	1,119	7	15	1,141	12	43
Nonfragile low income	350	52	1,235	1,637	17	22
Fragile low income	69	—	—	69	1	15
Total	2,905	3,673	3,215	9,793	100	14
Share of total stock (%)	30	38	33	100		
Share of total infrastructure stock (%)	4	5	4	14		

Source: Adapted from Irving and Manroth 2009.
Note: — = data not available.

2006, with 80 percent of this amount concentrated in South Africa (Irving and Manroth 2009). As of 2006, the power sector accounted for only 1.4 percent of the value of the syndicated infrastructure loans in Africa.

The two major power sector transactions based on syndicated loans for 2006 are reported in table 7.10. Much of this finance is denominated in local currency. Maturities are four to nine years in length with undisclosed spreads. The largest loan is the UNICEM power plant construction loan in Nigeria, which comprised a $210.6 million mixed naira-dollar–denominated loan delivered in four tranches raised from eight local banks, one U.S. bank (Citibank), and a local affiliate of a regional Ecobank.

Equity

Although the infrastructure companies issue only 7.7 percent of total value of corporate equities in the region, equity financing is a large part of overall local capital market infrastructure financing. A total of $55.9 billion of capital has been raised for infrastructure in this way, including $48.1 billion in South Africa alone and $7.8 billion outside South Africa (table 7.8). The region's stock exchanges played an important role in raising capital for the power sector, with $3.3 billion raised in this way in

Table 7.10 Syndicated Loan Transactions for Power Sector in 2006

Country	Borrower	Project	Loan amount ($ million)	Currency denomination	Number of tranches	Maturity	Pricing	Bank participation: local vs. nonlocal
Nigeria	UNICEM	Power plant construction	210.6	Naira and dollar	4	4 years, 7 years, 9 years	Undisclosed	8 local; 1 U.S. (Citibank); 1 local affiliate of regional Ecobank
Kenya	Iberafrica Power	Electric utility	16.8	Dollar	1	5 years	Undisclosed	1 local; Banque de Afrique (Benin); 1 local subsidiary of Stanbic Bank

Source: Adapted from Irving and Manroth 2009.

Africa overall, including $2.0 billion in South Africa and $1.3 billion outside South Africa (table 7.11).

As of 2006, the largest outstanding value was a KenGen issue on the Nairobi stock exchange that constituted 71 percent of total outstanding equity value in the power sector. The second largest equaled one-quarter of the total value in the sector. The remaining issues were quite small. Overall, power issues account for 2 percent of Sub-Saharan Africa's stock exchange listings by value (table 7.11).

Corporate Bonds

In the last decade, governments in the region extended the maturity profile of their security issues in an effort to establish a benchmark against which corporate bonds can be priced. Except in South Africa, however, corporate bond markets remain small and illiquid, where they exist at all. At 13 percent of GDP, South Africa's corporate bond market is by far the largest in the region, with $33.8 billion in issues outstanding at the end of 2006, followed by Namibia's at $457 million (7.1 percent of GDP). Outside South Africa, the few countries that had corporate bonds listed on their national or regional securities exchange at the end of 2006 had only a handful of such listings, and the amounts issued were small.

Overall, $3.7 billion of corporate bonds issued by power companies were outstanding as of the end of 2006 (table 7.8). As much as 98 percent of these were issued in South Africa by Eskom, which represents

Table 7.11   Details of Corporate Equity Issues by Power Sector Companies by End of 2006

Country	Issuer	Stock exchange	Outstanding value ($ million)	Percentage of all stocks on country exchange
Côte d'Ivoire	Compagnie Ivoirienne d'Electricitév	BRVM	53.4	4.0
Kenya	Kenya Power & Lighting Ltd.	Nairobi SE	307.9	2.7
	KenGen	Nairobi SE	926.6	8.0
	Kenya Power & Lighting Ltd. Pref. 4%	Nairobi SE AIM	0.2	0.002
	Kenya Power & Lighting Ltd. Pref. 7%	Nairobi SE AIM	0.05	0.0004
Nigeria	Nigeria Energy Sector Fund	Nigeria SE	14.8	0.06
Total electricity generation/power			1,302.9	2.0

Source: Adapted from Irving and Manroth 2009.
Note: AIM = alternative investment market; BRVM = Bourse Régionale des Valeurs Mobilières (regional stock exchange).

11 percent of the total value of outstanding corporate bonds and 53 percent of outstanding infrastructure bonds in that country. Only $0.5 billion in power sector bonds were issued outside South Africa in countries such as Benin, Burkina Faso, Kenya, Mozambique, Namibia, Senegal, Uganda, and Zambia. These small bond issues represent a large portion of total bond value in the respective countries. A single listing of Communauté Electrique de Benin in a small amount of $33.2 million accounted for 60 percent of total corporate bonds outstanding on BRVM. A listing of Zambia's Lunsemfwa Hydro Power in the amount of $7.0 million represented 43 percent of the Lusaka Stock Exchange's corporate bond value (table 7.12).

Institutional investors, including pension funds and insurance companies, could potentially become an important source of infrastructure financing in the future, with approximately $92 billion in assets accumulated in national pension funds and more than $181 billion in insurance assets. However, only a fraction of 1 percent of these assets is invested in infrastructure. It is not expected that this situation will change in the near future or without significant improvement in the macroeconomic environment.

Costs of Capital from Different Sources

The various sources of infrastructure finance reviewed in the previous sections differ greatly in their associated costs of capital (figure 7.7). For public funds, raising taxes is not a costless exercise. Each dollar raised and

Table 7.12   Details of Corporate Bonds Issued by Telecom Operators by End of 2006

Country	Issuer	Stock exchange	Issue date	Maturity terms (years)	Outstanding value ($ million)	Percentage of all corporate bond issues
Benin	Communauté Electrique de Benin	BRVM	2003	7	33.2	60
	Communauté Electrique de Benin	BRVM	2004	7	18.7	34
Zambia	Lunsemfwa Hydro Power	LuSE	2003	n.a.	7.0	43
Total electricity generation/power					58.9	6

Source: Adapted from Irving and Manroth 2009.
Note: BRVM = Bourse Régionale des Valeurs Mobilières (regional stock exchange); LuSE = Lusaka Stock Exchange; n.a. = not available.

Figure 7.7   Costs of Capital by Funding Source

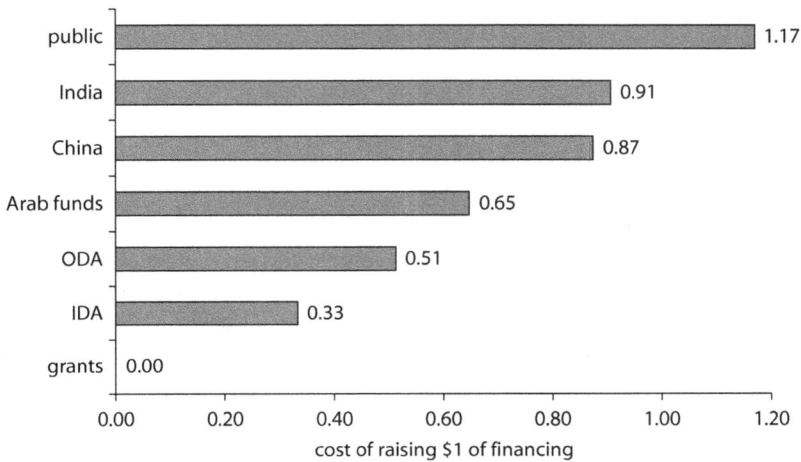

Source: Average marginal cost of public funds as estimated by Warlters and Auriol (2005); cost of equity for private sector as in Estache and Pinglo (2004) and Sirtaine and others (2005); authors' calculations.
Note: IDA = International Development Association; ODA = official development assistance.

spent by a Sub-Saharan African government has a social value premium (or marginal cost of public funds) of almost 20 percent. That premium captures the incidence of that tax on the society's welfare (caused by changes in consumption patterns and administrative costs, among other things). To allow ready comparisons across financing sources, this study standardized the financial terms as the present value of a dollar raised through each of the different sources. In doing so, it recognized that all loans must ultimately be repaid with tax dollars, each of which attracts the 20 percent cost premium.

Wide variation exists in lending terms. The most concessional IDA loans charge zero interest (0.75 percent service charge) with a 10-year grace period. India, China, and the Arab funds charge 4 percent, 3.6 percent, and 1.5 percent interest, respectively, and grant a four-year grace period.

The cost of non-OECD finance is somewhere between that of public funds and ODA. The subsidy factor for Indian and Chinese funds is about 25 percent and for the Arab funds, 50 percent. ODA typically provides a subsidy factor of 60 percent, rising to 75 percent for IDA resources. In addition to the cost of capital, sources of finance differ in the transaction costs associated with their use, which may offset or accentuate some of the differences.

The Most Promising Ways to Increase Funds

Given this setting, what are the best ways to increase availability of funds for infrastructure development? The place to start is clearly to get the most from existing budget envelopes by tackling inefficiencies. For some countries, this would be enough to close the funding gap in the power sector. For several others, however, particularly the fragile states, even after capturing all efficiency gains, a significant funding gap would remain. The prospects for improving this situation are not good, especially considering the long-term consequences of the recent financial crisis. The possibility exists across the board that all sources of infrastructure finance in Africa may fall rather than increase, which would further widen the funding gap. Only resource-rich countries have the possibility of using natural resource savings accounts to provide a source of financing for infrastructure, but only if macroeconomic conditions allow.

What Else Can Be Done?

The continent faces a substantial funding gap for power even if all the existing sources of funds—including efficiency gains—are tapped. What other options do these countries have? Realistically, they need either to defer the attainment of the infrastructure targets proposed here or to try to achieve them by using lower-cost technologies.

Taking More Time

The investment needs presented in this book are based on the objective of redressing Africa's infrastructure backlog within 10 years. It has been shown that it would be possible for middle-income states to meet this target within existing resource envelopes if the efficiency of resource use could be substantially improved. The same cannot be said for the other country groups. Extending the time horizon for the achievement of these goals should make the targets more affordable. But how long of a delay would be needed to make the infrastructure targets attainable without increasing existing spending envelopes?

By spreading the investment needs over 30 rather than 10 years, both resource-rich and nonfragile low-income countries could achieve the proposed targets within the existing spending envelopes. The fragile low-income countries would need to spread the investment needs over 60 years to reach the targets using the existing spending levels. These estimates are contingent on achieving efficiency gains, without which the time horizon for meeting the targets would be substantially longer than 30 and 60 years, respectively. Alternatively, the countries would need to considerably increase their existing spending to reach the targets (figure 7.8a).

Lowering Costs through Regional Integration

As we have already shown, regional integration is a crucial step in the power sector reform that would substantially reduce costs, mainly

Figure 7.8   Spreading Investment over Time

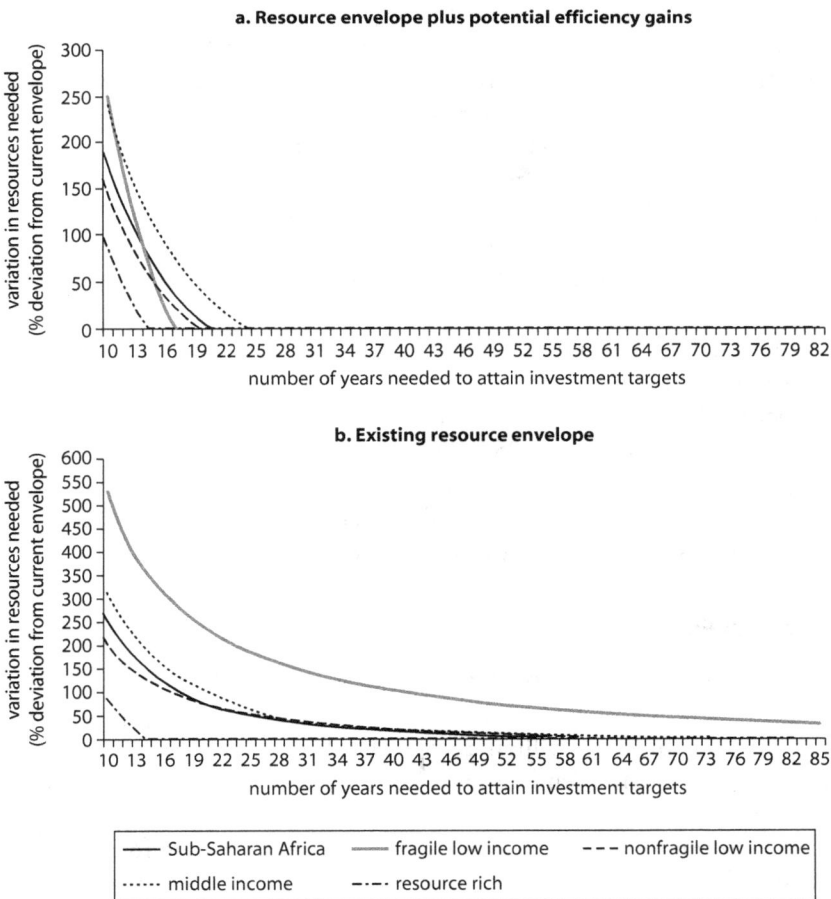

a. Resource envelope plus potential efficiency gains

y-axis: variation in resources needed (% deviation from current envelope)

x-axis: number of years needed to attain investment targets

b. Existing resource envelope

y-axis: variation in resources needed (% deviation from current envelope)

x-axis: number of years needed to attain investment targets

Legend:
—— Sub-Saharan Africa   —— fragile low income   – – – nonfragile low income
······ middle income   –·–· resource rich

Source: Foster and Briceño-Garmendia 2009.
Note: The threshold is the index value of 100.

because of economies of scale and increased share of hydropower in total power generation.

Pooling energy resources through regional power trade promises to significantly reduce power costs. In recognition of this benefit, regional power pools have been formed in Southern, West, East, and Central Africa and are at varying stages of maturity. If pursued to its full economic potential, regional trade could reduce the annual costs of power system operation and development by $2.7 billion (assuming efficiency gains have been achieved). The savings come largely from substituting hydropower for thermal power, which would lead to a substantial reduction in operating costs, even though it entails higher investment in capital-intensive hydropower and associated cross-border transmission in the short run. The returns to cross-border transmission can be as high as 120 percent (Southern African Power Pool) or more—typically 20–30 percent for the other power pools. By increasing the share of hydropower, regional trade would also save 70 million tons per year of carbon dioxide emissions.

Regional power trade would lead to an increase in the share of hydropower in Africa's generation portfolio from 36 percent to 48 percent, displacing 20,000 MW of thermal plant and saving 70 million tons per year of carbon dioxide emissions (8 percent of Sub-Saharan Africa's anticipated emissions through 2015).

Optimizing power trade would require 82 gigawatts (GW) of additional generation capacity and 22 GW of new cross-border transmission capacity. New generation, transmission, and distribution will require a substantial investment of $25 billion a year for the next 10 years, but the long-term marginal cost of producing and distributing power, which takes into account construction costs, still averages 13 percent below the current total costs and only 40 percent above the current effective tariffs.

The Way Forward

The cost of meeting the power sector spending needs estimated in this volume amounts to $40.1 billion a year, far above existing power sector spending of $11.6 billion a year. The difference between spending needs and current spending cannot be bridged entirely by capturing the estimated $8.2 billion a year of inefficiencies that exist at present, mainly in poorly operated utilities. No exceptions can be found to this general conclusion among country types: No country group covers more than 50 percent of its power sector funding gap by eliminating inefficiencies.

The inefficiencies in question arise from system losses, undercollection and overstaffing ($4.4 billion a year), underrecovery of costs ($3.6 billion a year), and underexecution of capital budgets ($0.2 billion a year). These findings underscore the importance of completing the reform agenda outlined previously to ensure adequate investment and O&M budgets.

Reforming public utilities and improving their operating performance will both increase the level of reinvestment from own resources and reduce their credit risk, enabling them to more easily access private debt markets. Policy and regulatory reforms are important for increased private sector participation.

Raising further finance for power infrastructure, particularly investment in new capacity and transmission, will be challenging. Historically, the main sources of finance have been public budgets and ODA, both of which are likely to suffer as a result of the 2008 global financial crisis. More emphasis will need to be placed on increasing finance from the private sector and non-OECD sources.

Note

1. For a detailed analysis of electricity tariffs and cost recovery issues in Sub-Saharan Africa, see Briceño-Garmendia and Shkaratan (2010).

References

Briceño-Garmendia, Cecilia, and Maria Shkaratan. 2010. "Power Tariffs: Caught between Cost Recovery and Affordability." Working Paper 20, Africa Infrastructure Country Diagnostic, World Bank, Washington, DC.

Briceño-Garmendia, Cecilia, Karlis Smits, and Vivien Foster. 2008. "Financing Public Infrastructure in Sub-Saharan Africa: Patterns and Emerging Issues." Background Paper 15, Africa Infrastructure Country Diagnostic, World Bank, Washington, DC.

Calderón, C., and L. Servén. 2004. "The Effects of Infrastructure Development on Growth and Income Distribution." Policy Research Working Paper 3400, World Bank, Washington, DC.

Collier, Paul, and Stephen O'Connell. 2006. "Opportunities and Choices in Explaining African Economic Growth." Centre for the Study of African Economies, Oxford University.

Estache, Antonio, and Maria Elena Pinglo. 2004. "Are Returns to Private Infrastructure in Developing Countries Consistent with Risks since the Asian Crisis?" Policy Research Working Paper 3373, World Bank, Washington, DC.

Foster, Vivien, and Cecilia Briceño-Garmendia, eds. 2009. *Africa's Infrastructure: A Time for Transformation.* Paris, France, and Washington, DC: Agence Française de Développement and World Bank.

Foster, Vivien, William Butterfield, Chuan Chen, and Nataliya Pushak. 2008. "Building Bridges: China's Growing Role as Infrastructure Financier for Sub-Saharan Africa." Trends and Policy Options No. 5, Public-Private Infrastructure Advisory Facility, World Bank, Washington, DC.

IMF (International Monetary Fund). 2007. *Regional Economic Outlook: Sub-Saharan Africa.* Washington, DC: IMF.

———. 2009. *The State of Public Finances: Outlook and Medium-Term Policies after the 2008 Crisis.* Washington, DC: IMF.

Irving, Jacqueline, and Astrid Manroth. 2009. "Local Sources of Financing for Infrastructure in Africa: A Cross-Country Analysis." Policy Research Working Paper 4878, World Bank, Washington, DC.

ODI (Overseas Development Institute). 2009. *A Development Charter for the G-20.* London: ODI.

PPIAF (Public-Private Infrastructure Advisory Facility). 2008. *Private Participation in Infrastructure Project Database.* Washington, DC: World Bank. http://ppi.worldbank.org/.

Pritchett, Lant. 1996. "Mind Your P's and Q's. The Cost of Public Investment Is Not the Value of Public Capital." Policy Research Working Paper 1660, World Bank, Washington, DC.

Rajkumar, Andrew, and Vinaya Swaroop. 2002. "Public Spending and Outcomes: Does Governance Matter?" Policy Research Working Paper 2840, World Bank, Washington, DC.

Reinikka, Ritva, and Jakob Svensson. 2002. "Explaining Leakage of Public Funds." Discussion Paper 3227, Centre for Economic Policy Research, London.

———. 2004. "The Power of Information: Evidence from a Newspaper Campaign to Reduce Capture." Policy Research Working Paper 3239, World Bank, Washington, DC.

Rosnes, Orvika, and Haakon Vennemo. 2008. "Powering Up: Costing Power Infrastructure Spending Needs in Sub-Saharan Africa." Background Paper 5, AICD, World Bank, Washington, DC.

Sirtaine, Sophie, Maria Elena Pinglo, J. Luis Guasch, and Viven Foster. 2005. "How Profitable Are Infrastructure Concessions in Latin America? Empirical Evidence and Regulatory Implications." Trends and Policy Options No. 2, Public-Private Infrastructure Advisory Facility, World Bank, Washington, DC.

UBS Investment Research. 2008. "Global Economic Perspectives: The Global Impact of Fiscal Policy." UBS, London.

Warlters, Michael, and Emmanuelle Auriol. 2005. "The Marginal Cost of Public Funds in Africa." Policy Research Working Paper 3679, World Bank, Washington, DC.

World Bank. 2005. *Global Monitoring Report 2005*. Washington, DC: World Bank.

———. 2007. DEPweb glossary. Development Education Program, World Bank, Washington, DC. http://www.worldbank.org/depweb/english/modules/glossary.html#middle-income.

World Economic Outlook. 2008. "Estimating the Size of the European Stimulus Packages for 2009." International Monetary Fund, Washington, DC.

APPENDIX 1

Africa Unplugged

Table A1.1 National Power System Characteristics

	Installed generation capacity (MW)	Installed generation capacity per million people (MW/million people)	Operational capacity as percentage of installed capacity (%)	System capacity factor (%)[a]	Generation technology, % of country total			
					Hydro	Oil	Gas	Coal
Benin	122	14	36.4	20.9	0.0	100.0	0.0	0.0
Botswana	132	70	100.0	62.8	0.0	17.0	0.0	83.0
Burkina Faso	180	13	100.0	32.7	24.6	75.0	0.0	0.0
Cameroon	902	54	100.0	50.3	92.0	8.0	0.0	0.0
Cape Verde	78	150	100.0	6.5	0.0	95.5	0.0	4.5
Chad	29	3	100.0	45.9	0.0	100.0	0.0	0.0
Congo, Dem. Rep.	2,443	41	40.9	82.1	96.4	3.6	0.0	0.0
Congo, Rep.	120	33	—	38.8	82.3	17.7	0.0	0.0
Côte d'Ivoire	1,084	59	100.0	58.2	42.3	11.4	45.8	0.0
Ethiopia	755	10	95.5	41.0	89.1	9.7	0.0	0.0
Ghana	1,622	72	100.0	44.5	53.3	34.3	6.7	0.0
Kenya	1,211	34	87.6	57.5	58.1	27.2	0.0	0.0
Lesotho	76	42	95.1	65.0	97.9	2.1	0.0	0.0
Madagascar	227	12	61.7	79.3	56.8	43.2	0.0	0.0
Malawi	285	22	91.6	59.8	91.4	3.2	0.0	0.0
Mali	278	23	79.0	38.7	56.3	43.7	0.0	0.0
Mozambique	233	12	63.5	—	93.8	5.6	0.0	0.0
Namibia	393	192	100.0	45.9	62.2	6.6	0.0	31.1

Niger	105	87.9	25.0	0.0	69.1	0.0	30.9
Nigeria	5,898	63.4	73.6	22.8	14.8	60.3	0.0
Rwanda	31	100.0	42.8	95.6	—	0.0	0.0
Senegal	300	100.0	65.8	0.0	96.1	0.0	0.0
South Africa	40,481	89.9	71.5	5.5	2.0	0.3	86.7
Sudan	801	100.0	—	18.6	66.0	9.4	0.0
Tanzania	881	96.3	25.3	52.5	24.8	17.7	0.6
Uganda	321	70.2	—	85.3	11.8	0.0	0.0
Zambia	1,778	66.6	85.3	92.9	5.3	0.0	0.0
Zimbabwe	1,960	98.5	43.5	38.3	0.0	0.0	61.7
Benchmarks (Weighted Averages)							
Sub-Saharan Africa	4,760	85.1	68.1	18.1	6.0	5.9	65.4
CAPP	709	100.0	50.1	77.7	22.3	0.0	0.0
EAPP	1,169	68.9	63.0	79.0	14.0	1.8	0.1
SAPP	9,855	86.4	71.2	13.2	2.5	0.1	79.4
WAPP	3,912	76.0	64.0	29.4	23.4	44.4	0.2
Predominantly thermal capacity	9,129	86.7	71.3	7.8	4.4	5.6	77.5
Predominantly hydro capacity	1,101	79.5	60.8	73.7	14.8	7.0	1.1
Installed capacity high	7,625	84.7	71.0	15.2	4.6	6.1	69.4
Installed capacity medium	597	93.6	49.0	65.2	27.9	2.0	3.0
Installed capacity low	73	78.6	30.1	52.9	40.0	0.0	6.4

Source: Eberhard and others 2008.

Note: Data as of 2005 or the earliest year available before 2005. For Botswana, Republic of Congo, Mali, and Zimbabwe, data for 2007. kWh = kilowatt hour; MW = megawatt; CAPP = Central African Power Pool; EAPP = East African Power Pool; SAPP = Southern African Power Pool; WAPP = West African Power Pool.

— = data not available.

a. Calculated as ratio of electricity generated (watt-hours [Wh]) to installed operational capacity in Wh (W x 24 x 365).

Table A1.2   Electricity Production and Consumption

	Generation and net imports			Consumption of electricity by customer group, as percentage of total (%)			Consumption per customer by customer group (kWh/month)		
	Net electricity generated (GWh/year)	Net electricity generated per capita (kWh/capita/year)	Net import (import-export) (GWh/year)	Low voltage	Medium voltage	High voltage	Low voltage	Medium voltage	High voltage
Benin	81	9	595	40.1	40.1	19.8	35	22,079	—
Botswana	726	386	2,394	—	—	—	—	—	—
Burkina Faso	516	38	0	63.1	36.9	0.0	111	22,141	—
Cameroon	4,004	240	0	32.8	22.9	44.2	170	49,537	40,108,583
Cape Verde	45	87	0	49.7	38.0	12.3	94	2,266	—
Chad	117	12	0	63.5	36.5	0.0	205	33,904	—
Congo, Dem. Rep.	7,193	121	-1,794	0.0	14.7	85.3	—	—	31,397,692
Congo, Rep.	407	115	449	—	—	—	—	—	—
Côte d'Ivoire	5,524	299	-1,397	51.3	48.7	0.0	147	47,018	—
Ethiopia	2,589	36	0	34.9	25.2	39.9	77	387	550,933
Ghana	6,750	300	176	13.4	10.1	76.5	71	1,949	5,214,850
Kenya	5,347	152	28	34.1	20.4	45.4	169	22,997	366,611
Lesotho	410	229	13	38.2	35.1	26.7	293	3,068	54,278
Madagascar	973	51	0	58.6	36.8	4.6	92	25,610	962,978
Malawi	1,368	104	0	35.7	17.1	47.2	252	496	61,300
Mali	942	76	2.27	—	—	—	—	—	—
Mozambique	147	7	-2,413	36.8	40.9	22.3	127	28,998	—
Namibia	1,580	770	1,489	3.5	4.0	92.5	161	3,068	—
Niger	202	14	220	60.6	0.1	39.4	134	20	5,466,667
Nigeria	24,079	166	0	51.0	27.0	22.0	897	739	22,692

Rwanda	116	13	110	—	—	—	—	—	4,805,556
Senegal	2,105	176	0	58.5	31.3	10.1	139	39,774	1,149,338
South Africa	228,071	4,813	−2,343	7.9	5.8	86.3	336	12,170	—
Sudan	—	—	—	—	—	—	—	—	—
Tanzania	1,880	48	136	43.6	14.9	41.5	162	24,898	169,305
Uganda	1,893	63	−170	19.0	7.6	73.4	120	—	150,682
Zambia	8,850	746	222	—	—	—	—	—	—
Zimbabwe	7,471	557	2,659	—	—	—	—	—	—
Benchmarks (Weighted Averages)									
Sub-Saharan Africa	23,337	470	−214	32	22	33	2,049	81,492	4,285,319
CAPP	2,583	147	0	31	21	39	2,051	584,560	481,303,000
EAPP	3,808	77	−282	21	17	62	1,591	18,972	2,483,018
SAPP	52,890	1,214	−521	7	6	80	2,161	113,767	15,830,794
WAPP	16,079	171	−51	44	27	27	2,234	90,760	370,686
Predominantly thermal	53,340	963	−153	12	8	77	2,360	20,203	4,454,510
Predominantly hydro	3,925	161	−257	21	17	41	1,694	281,735	2,968,714
Installed capacity high	41,975	822	−836	12	9	74	2,355	96,812	4,523,681
Installed capacity medium	2,113	77	162	33	19	42	1,561	17,773	1,978,348
Installed capacity low	149	22	156	28	28	8	1,058	49,345	855,338

Source: Eberhard and others 2008.

Note: Data as of 2005 or the earliest year available before 2005. For Botswana, Republic of Congo, Mali, and Zimbabwe, data as of 2007. GW = gigawatt; kWh = kilowatt hour; CAPP = Central African Power Pool; EAPP = East African Power Pool; SAPP = Southern African Power Pool; WAPP = West African Power Pool.

— = data not available.

Table A1.3 Outages and Own Generation: Statistics from the Enterprise Survey

	Electricity cited as business constraint, % firms	Power outages (days)	Power outages, % sales	Equipment destroyed by outages, % sales	Generator owners, % firms	Power from own generator (%)
Algeria	11.47	12.32	5.28	—	29.49	6.22
Benin	69.23	56.12	7.79	1.54	26.90	32.80
Botswana	9.65	22.29	1.54	—	14.91	17.57
Burkina Faso	68.97	7.82	3.87	—	29.82	6.52
Burundi	79.56	143.76	11.75	—	39.22	25.28
Cameroon	64.94	15.80	4.92	—	57.79	7.62
Cape Verde	70.69	15.18	6.87	—	43.10	13.53
Egypt, Arab Rep.	26.46	10.40	6.12	—	19.26	5.87
Eritrea	37.66	74.61	5.95	—	43.04	9.31
Ethiopia	42.45	44.16	5.44	—	17.14	1.58
Kenya	48.15	53.40	9.35	0.34	73.40	15.16
Madagascar	41.30	54.31	7.92	0.91	21.50	2.23
Malawi	60.38	63.21	22.64	—	49.06	4.44
Mali	24.18	5.97	2.67	1.36	45.33	5.09
Mauritania	29.66	37.97	2.06	—	26.25	11.75
Mauritius	12.68	5.36	4.01	0.42	39.51	2.87
Morocco	8.94	3.85	0.82	—	13.81	11.16
Namibia	15.09	0.00	1.20	—	13.21	13.33
Niger	26.09	3.93	2.72	—	27.54	14.74
Senegal	30.65	25.64	5.12	0.62	62.45	6.71
South Africa	8.96	5.45	0.92	—	9.45	0.17
Swaziland	21.43	32.38	1.98	—	35.71	10.33
Tanzania	60.24	63.09	—	0.81	59.05	12.28
Uganda	43.85	45.50	6.06	0.74	38.34	6.56
Zambia	39.61	25.87	4.54	0.28	38.16	5.13
Average	38.09	33.14	5.48	0.78	34.94	9.93

Source: Foster and Steinbuks 2008.
Note: — = data not available.

Table A1.4   Emergency, Short-Term, Leased Generation

	Emergency generation capacity (MW)	Total generation capacity (MW)	Emergency generation capacity, % of total	Cost of emergency generation, % GDP
Sierra Leone	20	49	40.8	4.25
Uganda	100	303	33.0	3.29
Madagascar	50	227	22.0	2.79
Ghana	80	1,490	5.4	1.90
Rwanda	15	39	38.5	1.84
Kenya	100	1,211	8.3	1.45
Senegal	40	428	9.3	1.37
Angola	150	830	18.1	1.04
Tanzania	40	881	4.5	0.96
Gabon	14	414	3.4	0.45
Total	609	5,872	183.0	19.00

Source: Foster and Steinbuks 2008.
Note: Emergency power plant is generally leased for short periods and thus the amount of emergency power in individual countries varies from year to year. GDP = gross domestic product; MW = megawatt.

Table A1.5   Distribution of Installed Electrical Generating Capacity between Network and Private Sector Self-Generation

	Utility and government, % of total	Self-generation, by sector, % of total			
		Mining	Fuels	Manufacturing	Commerce/ services
Angola	94	2.69	2.65	0.19	0.14
Benin	97	0.00	0.00	3.13	0.00
Botswana	91	8.84	0.00	0.00	0.00
Burkina Faso	100	0.00	0.00	0.00	0.00
Burundi	98	0.00	0.00	0.41	1.84
Cameroon	99	0.00	0.95	0.00	0.01
Cape Verde	100	0.00	0.00	0.00	0.09
Central African Republic	100	0.00	0.00	0.00	0.23
Chad	100	0.00	0.00	0.00	0.00
Comoros	100	0.00	0.00	0.00	0.00
Congo, Dem. Rep.	93	3.19	1.18	0.13	2.55
Congo, Rep.	53	0.00	42.39	4.43	0.00
Côte d'Ivoire	96	0.00	3.41	0.53	1.48
Equatorial Guinea	49	0.00	51.34	0.00	0.00
Eritrea	100	0.00	0.00	0.00	0.00
Ethiopia	99	0.00	0.00	0.17	0.36

(continued next page)

Table A1.5   *(continued)*

	Utility and government, % of total	Self-generation, by sector, % of total			
		Mining	Fuels	Manufacturing	Commerce/ services
Gabon	81	0.76	18.15	0.00	0.00
Gambia	98	0.00	0.00	0.00	1.55
Ghana	87	0.31	12.53	0.23	0.00
Guinea	82	18.49	0.00	0.00	0.00
Guinea-Bissau	94	0.00	0.00	6.43	0.00
Kenya	95	0.00	0.00	4.28	0.68
Lesotho	100	0.00	0.00	0.00	0.00
Liberia	100	0.00	0.00	0.00	0.00
Madagascar	100	0.00	0.00	0.00	0.02
Malawi	94	0.00	0.00	5.92	0.00
Mali	91	8.14	0.00	0.00	0.69
Mauritania	38	57.10	4.19	0.00	0.23
Mauritius	77	0.00	0.00	20.31	3.03
Mozambique	99	0.00	0.00	0.81	0.33
Namibia	100	0.00	0.00	0.00	0.00
Niger	76	21.79	3.45	1.89	0.00
Nigeria	78	0.50	10.84	3.09	7.75
Rwanda	93	0.00	0.00	4.98	1.83
São Tomé and Príncipe	100	0.00	0.00	0.00	0.00
Senegal	99	0.00	0.00	0.00	0.73
Seychelles	100	0.00	0.00	0.00	0.00
Sierra Leone	100	0.00	0.00	0.00	0.00
Somalia	100	0.00	0.00	0.00	0.00
South Africa	98	0.22	0.63	0.92	0.03
Sudan	93	0.00	1.32	5.31	0.02
Swaziland	49	2.97	0.00	48.53	0.00
Tanzania	89	5.32	0.56	3.58	1.99
Togo	79	20.50	0.00	0.00	1.00
Uganda	93	3.70	0.00	0.32	3.39
Zambia	98	0.00	0.00	0.68	1.38
Zimbabwe	96	0.00	0.00	4.11	0.02
Average	90	3.29	3.27	2.56	0.67

Source: Foster and Steinbuks 2008.

Table A1.6   Effect of Own Generation on Marginal Cost of Electricity
$/kWh

	Average variable cost of own electricity (A)	Average capital cost of own electricity (B)	Average total cost of own electricity (C = A + B)	Price of kWh purchased from public grid (D)	Weighted average cost of electricity (E = ∂C+[1−∂]D)[d]
Algeria	0.04	0.11	0.15	0.03[c]	0.05
Benin	0.36	0.10	0.46	0.12[c]	0.27
Burkina Faso[a]	0.42	0.32	0.74	0.21[c]	0.23
Cameroon[a]	0.41	0.04	0.46	0.12[c]	0.16
Cape Verde[a]	0.46	0.04	0.50	0.17[c]	0.26
Egypt, Arab Rep.	0.04	0.26	0.30	0.04[c]	0.12
Eritrea	0.11	0.03	0.13	0.11	0.12
Kenya	0.24	0.06	0.29	0.10	0.14
Madagascar	0.31	0.08	0.39	—	—
Malawi	0.46	0.03	0.50	0.05[c]	0.09
Mali	0.26	0.26	0.52	0.17	0.21
Mauritius	0.26	0.35	0.61	0.14[c]	0.25
Morocco	0.31	0.32	0.62	0.08[c]	0.15
Niger[a]	0.36	0.04	0.41	0.23[c]	0.26
Senegal	0.25	0.09	0.34	0.16	0.18
Senegal[b]	0.28	0.40	0.68	0.16	0.30
South Africa	0.18	0.36	0.54	0.04	0.05
Tanzania	0.25	0.04	0.29	0.09	0.13
Uganda	0.35	0.09	0.44	0.09	0.14
Zambia	0.27	0.18	0.45	0.04	0.06
Average	0.28	0.15	0.43	0.11	0.16

Source: Foster and Steinbuks 2008.
Note: kWh = kilowatt hour; — = data not available.
a. Tourism industry (hotels and restaurants sector) only.
b. Survey of informal sector.
c. Data not reported in the enterprise surveys (obtained from the public utilities).
d. ∂ Share of total electricity consumption coming from own generation.

Table A1.7   Losses Due to Outages ("Lost Load") for Firms with and without Their Own Generator
$/hour

	Without own generator	With own generator
Algeria	155.8	52.2
Benin	38.4	23.1
Burkina Faso[a]	114.1	13.0
Cameroon[a]	403.6	12.3

(continued next page)

Table A1.7   *(continued)*

	Without own generator	*With own generator*
Cape Verde[a]	177.7	36.4
Egypt, Arab Rep.	201.5	30.4
Eritrea	31.9	10.2
Kenya	113.1	37.1
Madagascar	434.5	153.0
Malawi	917.3	401.4
Mali	390.3	9.5
Mauritius	468.6	13.9
Morocco	377.5	22.9
Niger[a]	81.3	22.6
Senegal	166.0	19.2
Senegal[b]	12.9	1.9
South Africa	1140.1	66.1
Tanzania	—	444.3
Uganda	27.6	191.4
Zambia	286.6	39.2
Average	307.0	84.1

Source: Foster and Steinbuks 2008.
Note: — = data not available.
a. Survey of tourism sector.
b. Survey of informal sector.

Table A1.8   Operating Costs of Own Generation

	Fuel price (cents/liter)	*Cost ($/kWh)*				
		<5 kVA	*5–100 kVA*	*100 kVA–1 MW*	*1 MW–10 MW*	*Grid*
Algeria	0.10	0.08	0.05	0.03	0.03	0.03
Benin	0.72	0.58	0.32	0.22	0.19	0.12
Botswana	0.61	0.49	0.27	0.18	0.16	0.04
Burkina Faso	0.94	0.75	0.42	0.28	0.25	0.21
Burundi	1.08	0.86	0.49	0.32	0.29	—
Cameroon	0.83	0.66	0.37	0.25	0.22	0.12
Cape Verde	0.81	0.65	0.36	0.24	0.22	0.17
Egypt, Arab Rep.	0.10	0.08	0.05	0.03	0.03	0.04
Eritrea	0.25	0.20	0.11	0.08	0.07	0.11
Ethiopia	0.32	0.26	0.14	0.10	0.09	0.06
Kenya	0.56	0.45	0.25	0.17	0.15	0.10
Madagascar	0.79	0.63	0.36	0.24	0.21	—
Malawi	0.88	0.70	0.40	0.26	0.24	0.05
Mali	0.55	0.44	0.25	0.17	0.15	0.17

(continued next page)

Table A1.8   *(continued)*

	Fuel price (cents/liter)	*Cost ($/kWh)*				
		<5 kVA	*5–100 kVA*	*100 kVA–1 MW*	*1 MW–10 MW*	*Grid*
Mauritania	0.59	0.47	0.27	0.18	0.16	—
Mauritius	0.56	0.45	0.25	0.17	0.15	0.14
Morocco	0.70	0.56	0.32	0.21	0.19	0.08
Namibia	0.65	0.52	0.29	0.20	0.18	0.04
Niger	0.91	0.73	0.41	0.27	0.25	0.23
Senegal	0.53	0.42	0.24	0.16	0.14	0.16
South Africa	0.40	0.32	0.18	0.12	0.11	0.04
Swaziland	0.73	0.58	0.33	0.22	0.20	0.05
Tanzania	0.61	0.49	0.27	0.18	0.16	0.09
Uganda	0.70	0.56	0.32	0.21	0.19	0.09
Zambia	0.60	0.48	0.27	0.18	0.16	0.04
Average	0.62	0.50	0.28	0.19	0.17	0.10

Source: Foster and Steinbuks 2008.
Note: kVA = kilovolt-ampere; KWh = kilowatt hour; MW = megawatt; — = data not available.

The Promise of Regional Power Trade

Table A2.1   Projected Trading Patterns in 10 Years under Alternative Trading Scenarios, by Region

	Demand 2005 (TWh)	Demand 2015 (projected) (TWh)	Trade expansion, 2015				Trade stagnation, 2015			
			Imports (TWh)	Exports (TWh)	Net exports (TWh)	Trade, % of demand	Imports (TWh)	Exports (TWh)	Net exports (TWh)	Trade, % of demand
SAPP										
Angola	2.1	7.9	11.0	5.0	−6.0	65	0.1	0.1	0	0
Botswana	2.4	4.2	10.4	6.2	−4.3	93	1.2	6.5	5.3	−117
Congo, Dem. Rep.	4.7	13.6	0.4	52.2	51.9	−369	0.4	2.7	2.3	−16
Lesotho	0.4	0.9	0.7	0	−0.7	68	0.7	0	−0.7	68
Malawi	1.3	2.3	1.8	0.3	−1.5	56	0	1.6	1.6	−60
Mozambique	11.2	16.4	13.2	19.1	5.9	−33	7.0	9.2	2.1	−12
Namibia	2.6	4.3	4.9	1.1	−3.8	72	1.2	2.0	0.9	−17
South Africa	215.0	319.2	37.7	1.4	−36.4	10	21.8	7.8	−14.0	4.0
Zambia	6.3	9.3	41.1	39.3	−1.8	18	2.4	8.4	6.0	−62
Zimbabwe	12.8	18.7	25.5	22.0	−3.5	17	12.4	8.8	−3.6	18
Total SAPP	**258.8**	**396.8**	**146.7**	**146.6**	**−0.2**	**−3**	**47.2**	**47.1**	**−0.1**	**−194**
EAPP/Nile Basin										
Burundi	0.2	0.7	0.9	0.2	−0.7	78	0	0	0	0
Ethiopia	2.1	10.7	3.4	29.6	26.2	−227	3.4	3.3	−0.1	1
Kenya	4.6	12.0	3.3	0.5	−2.8	22	1.0	0.5	−0.5	3
Sudan	3.2	9.2	29.2	42.4	13.1	−134	0.2	0.2	0	0
Tanzania	4.2	7.9	0.5	2.9	2.4	−22	0.5	0.5	0	0
Uganda	1.6	4.2	0.8	3.6	2.8	−61	0.4	0.8	0.4	−9
Total EAPP/ Nile Basin	**15.9**	**44.7**	**38.1**	**79.2**	**41.0**	**−344**	**5.5**	**5.3**	**−0.2**	**−5**

WAPP

Benin	0.6	1.7	1.0	0	-0.9	45	0.4	0	-0.4	18
Burkina Faso	0.5	1.5	1.0	0	-1.0	58	1.0	0	-1.0	58
Cape Verde	0.04	0.1	0	0	0	—	—	—	—	—
Côte d'Ivoire	2.9	5.4	11.1	12.0	0.9	-12	0.2	3.7	3.5	-47
Gambia, The	0.1	0.4	0	0.1	0.1	-19	0	0	0	0
Ghana	5.9	12.8	11.0	1.4	-9.6	52	2.8	0.3	-2.6	14
Guinea	0.7	2.1	0	17.4	17.4	-564	0	0	0	0
Guinea-Bissau	0.1	0.2	2.1	1.9	-0.2	77	0	0	0	0
Liberia	0.3	1.3	1.7	0	-1.7	89	0	0	0	0
Mali	0.4	1.8	12.7	10.9	-1.9	79	0.1	0.4	0.3	-14
Niger	0.4	1.3	1.5	0	-1.5	86	0.4	0	-0.4	20
Nigeria	16.9	59.2	2.1	4.2	2.1	-3	2.1	2.4	0.3	0
Senegal	1.5	3.5	2.0	0.7	-1.4	30	0.7	0.1	-0.6	13
Sierra Leone	0.2	1.0	2.6	1.7	-0.9	60	0	0	0	0
Togo	0.6	1.5	1.1	0.2	-0.9	48	0	0.5	0.5	-27
Total WAPP	**31.14**	**93.8**	**49.9**	**50.5**	**0.5**	**26**	**7.7**	**7.4**	**-0.4**	**35**

CAPP

Cameroon	3.4	6.4	0.2	6.9	6.7	-84	0.2	0.2	0	0
Central African Republic	0.1	0.5	0	0	0	0	0	0	0	0
Chad	0.1	0.9	1.3	0	-1.3	102	0	0	0	0
Congo, Rep.	5.8	10.3	4.4	0	-4.4	34	0	0	0	0
Equatorial Guinea	0.03	0.1	0.1	0	-0.1	100	0	0	0	0
Gabon	1.2	1.8	1.0	0	-1.0	42	0	0	0	0
Total CAPP	**10.63**	**20.0**	**7.0**	**6.9**	**-0.1**	**194**	**0.2**	**0.2**	**0.0**	**0.0**

Source: Rosnes and Vennemo 2008.
Note: CAPP = Central African Power Pool; EAPP/Nile Basin = East African/Nile Basin Power Pool; SAPP = Southern African Power Pool; WAPP = West African Power Pool; TWh = terawatt-hour.
— = data not available.

Table A2.2   Projected Long-Run Marginal Cost in 10 Years under Alternative Trading Scenarios

a. SAPP
cents/kWh

	Trade-expansion scenario			Trade-stagnation scenario		
	Cost of generation and international transmission lines	Cost of domestic T&D	Total LRMCs	Cost of generation and international transmission lines	Cost of domestic T&D	Total LRMCs
Angola	1.8	4.6	6.4	5.9	4.6	10.5
Botswana	3.5	2.1	5.6	4.0	2.1	6.1
Congo, Dem. Rep.	1.4	2.4	3.9	1.1	2.4	3.6
Lesotho	3.6	2.2	5.8	4.8	2.2	7.0
Malawi	3.3	1.9	5.1	3.5	1.9	5.4
Mozambique	3.3	0.8	4.1	4.7	0.8	5.5
Namibia	3.6	7.3	10.9	4.7	7.3	12.0
South Africa	3.6	2.0	5.5	4.7	2.0	6.7
Zambia	2.9	4.6	7.5	3.3	4.6	7.8
Zimbabwe	3.2	4.6	7.8	3.9	4.6	8.5
Average	**3.3**	**2.7**	**6.0**	**4.6**	**2.7**	**7.3**

b. EAPP/Nile Basin

cents/kWh

	Trade-expansion scenario			Trade-stagnation scenario		
	Cost of generation and international transmission lines	Cost of domestic T&D	Total LRMC	Cost of generation and international transmission lines	Cost of domestic T&D	Total LRMC
Burundi	6.8	4.6	11.4	10.3	4.6	14.9
Djibouti	6.6	0.6	7.2	6.6	0.6	7.2
Egypt, Arab Rep.	7.6	0.9	8.5	7.7	0.9	8.6
Ethiopia	6.9	12.2	19.0	4.0	12.2	16.1
Kenya	7.4	5.0	12.4	8.4	5.0	13.3
Rwanda	6.4	6.1	12.4	6.0	6.1	12.1
Sudan	7.5	5.2	12.7	7.4	5.2	12.6
Tanzania	6.5	3.2	9.7	4.5	3.2	7.8
Uganda	6.8	5.4	12.3	5.9	5.4	11.3
Average	**7.4**	**4.7**	**12.1**	**7.5**	**4.7**	**12.2**

(continued next page)

Table A2.2 *(continued)*

c. WAPP
cents/kWh

	Trade-expansion scenario			Trade-stagnation scenario		
	Cost of generation and international transmission lines	Cost of domestic T&D	Total LRMCs	Cost of generation and international transmission lines	Cost of domestic T&D	Total LRMCs
Benin	7.9	11.1	19.0	8.1	11.1	19.2
Burkina Faso	7.2	18.1	25.3	7.9	18.1	26.0
Côte d'Ivoire	6.9	7.8	14.7	7.6	7.8	15.4
Gambia, The	6.3	1.7	8.0	5.8	1.7	7.4
Ghana	7.3	2.3	9.6	8.0	2.3	10.3
Guinea	5.8	1.3	7.0	4.7	1.3	6.0
Guinea-Bissau	6.3	2.2	8.5	13.4	2.2	15.6
Liberia	6.6	1.5	8.1	12.6	1.5	14.1
Mali	6.4	18.2	24.6	9.7	18.2	27.9
Mauritania	6.9	6.7	13.6	7.8	6.7	14.5
Niger	7.9	16.8	24.7	13.6	16.8	30.4
Nigeria	7.6	5.2	12.8	7.6	5.2	12.8
Senegal	6.6	36.8	43.4	10.0	36.8	46.8
Sierra Leone	6.1	2.4	8.6	7.3	2.4	9.7
Togo	7.6	2.7	10.3	8.0	2.7	10.6
Average	**7.2**	**11.1**	**18.3**	**8.0**	**11.1**	**19.1**

d. CAPP
cents/kWh

	Trade-expansion scenario			Trade-stagnation scenario		
	Cost of generation and international transmission lines	Cost of domestic T&D	Total LRMC	Cost of generation and international transmission lines	Cost of domestic T&D	Total LRMC
Cameroon	4.4	2.4	6.9	4.0	2.4	6.4
Central African Republic	4.8	6.3	11.1	4.8	6.3	11.1
Chad	4.9	2.0	6.8	9.0	2.0	10.9
Congo, Rep.	5.4	0.2	5.6	7.9	0.2	8.1
Equatorial Guinea	4.8	2.7	7.5	7.0	2.7	9.7
Gabon	4.9	1.6	6.5	5.9	1.6	7.4
Average	**4.9**	**2.2**	**7.0**	**7.0**	**2.2**	**9.1**

Source: Rosnes and Vennemo 2008.

Note: Average is weighted by annualized cost. In some cases power exporting countries report higher LRMC under trade expansion. Even if the cost of meeting domestic power consumption may be higher with trade than without; the higher revenues earned from exports would more than compensate for that increment. CAPP = Central African Power Pool; kWh = kilowatt-hour; LRMC = long-run marginal cost; T&D = transmission and distribution.

Table A2.3 Projected Composition of Generation Portfolio in 10 Years under Alternative Trading Scenarios
% of total

	Trade expansion, 2015			Trade stagnation, 2015		
	Hydro capacity	Coal and gas	Other capacity	Hydro capacity	Coal and gas	Other capacity
Angola	88	4	8	80	16	3
Benin	16	0	84	12	28	61
Botswana	0	96	4	0	100	0
Burkina Faso	39	0	61	39	0	61
Burundi	43	0	57	60	0	40
Cameroon	95	0	5	91	0	9
Cape Verde	0	4	96	0	0	0
Central African Republic	100	0	0	100	0	0
Chad	0	0	100	0	0	100
Congo, Dem. Rep	100	0	0	99	0	1
Congo, Rep.	100	0	0	85	0	15
Côte d'Ivoire	77	20	2	79	19	2
Equatorial Guinea	0	100	0	73	27	0
Ethiopia	94	0	6	81	0	19
Gabon	99	0	1	65	0	35
Gambia, The	0	0	100	0	0	100
Ghana	75	19	7	67	28	6
Guinea	99	0	1	94	0	6
Guinea-Bissau	0	0	100	44	0	56
Kenya	42	39	19	36	48	17
Lesotho	95	0	5	95	0	5
Liberia	100	0	0	98	0	2
Madagascar	16	0	84	0	0	0

Country						
Malawi	91	6	3	98	2	1
Mali	73	6	21	82	3	14
Mauritania	0	0	100	0	58	42
Mauritius	30	60	9	0	0	0
Mozambique	97	3	0	69	30	0
Namibia	66	33	1	31	68	0
Niger	0	99	1	73	14	13
Nigeria	76	21	3	77	20	3
South Africa	8	82	9	8	83	9
Senegal	0	42	58	9	38	53
Sierra Leone	46	0	54	77	0	23
Sudan	87	6	7	73	13	14
Tanzania	61	35	4	67	29	4
Togo	99	0	1	56	44	0
Uganda	91	0	9	87	0	13
Zambia	95	4	1	97	2	1
Zimbabwe	73	26	1	47	52	1

Source: Rosnes and Vennemo 2008.

Table A2.4   Projected Physical Infrastructure Requirements in 10 Years under Alternative Trading Scenarios
MW

	Trade expansion				Trade stagnation			
	Generation capacity: installed	Generation capacity: refurbishment	Generation capacity: new investments	New cross-border transmission	Generation capacity: installed	Generation capacity: refurbishment	Generation capacity: new investments	New cross-border transmission
Angola	620	203	8	2,120	620	305	1,184	0
Benin	75	112	4	160	75	112	76	0
Botswana	120	12	2,141	2,120	120	12	0	0
Burkina Faso	66	95	0	0	66	95	0	0
Burundi	13	18	30	78	13	18	173	0
Cameroon	249	596	2,471	831	249	596	1,015	0
Cape Verde	0	1	18	0				
Central African Republic	0	18	143	0	0	18	143	0
Chad	0	0	0	202	0	22	174	0
Congo, Dem. Rep.	0	2,236	8,401	5,984	0	2,236	761	0
Congo, Rep.	0	129	1,689	498	0	148	2,381	0
Côte d'Ivoire	498	601	1,368	2,226	498	601	1,562	0
Equatorial Guinea	10	0	0	20	10	0	28	0
Ethiopia	831	335	8,699	2,997	831	335	1,933	0
Gabon	0	163	92	111	0	297	92	0
Gambia, The	22	74	4	19	22	66	4	0
Ghana	1,713	160	1,048	979	1,713	160	1,400	0
Guinea	119	28	4,288	2,283	119	28	568	0
Guinea-Bissau	0	22	0	818	0	22	24	0
Kenya	695	389	987	266	695	395	1,237	0

Lesotho	75	0	0	75	0	0	0
Liberia	0	0	258	0	73	369	0
Madagascar	87	349	0	—	—	—	—
Malawi	175	1	227	175	100	698	0
Mali	288	3	2,703	288	66	284	0
Mauritania	37	1	79	37	63	136	0
Mauritius	195	0	0	—	—	—	—
Mozambique	2,174	3,248	1,400	2,174	180	2,348	0
Namibia	240	2	556	240	120	1,059	0
Niger	0	0	206	0	73	202	0
Nigeria	1,011	10,828	366	1,011	5,208	10,828	0
South Africa	13,463	19,399	547	13,463	21,690	20,873	0
Senegal	205	258	487	205	139	320	0
Sierra Leone	62	1	661	62	46	145	0
Sudan	2,554	3,704	13,491	2,554	0	568	0
Tanzania	468	2046	266	468	305	1,767	0
Togo	0	200	5	0	91	321	0
Uganda	358	1,258	537	358	191	707	0
Zambia	268	3	7,526	268	1,670	1,726	0
Zimbabwe	0	2,251	3,072	0	1,835	2,158	0
SSA: Total	**26,654**	**74,942**	**54,020**	**26,372**	**37,253**	**57,128**	**0**
CAPP: Total	**259**	**4,395**	**1,662**	**259**	**1,081**	**3,833**	**0**
EAPP/Nile							
Basin: Total	**4,919**	**16,724**	**17,635**	**4,919**	**1,244**	**6,385**	**0**
SAPP: Total	**17,135**	**35,454**	**23,552**	**17,135**	**28,148**	**30,807**	**0**
WAPP: Total	**4,059**	**18,020**	**11,171**	**4,059**	**6,780**	**16,103**	**0**

Source: Rosnes and Vennemo 2008.

Note: CAPP = Central African Power Pool; EAPP/Nile Basin = East African/Nile Basin Power Pool; SAPP = Southern African Power Pool; SSA = Sub-Saharan Africa; WAPP = West African Power Pool; MW = megawatt. — = data not available.

Table A2.5 Estimated Annualized 10-Year Spending Needs to Meet Infrastructure Requirements under Alternative Trading Scenarios
$million/year

	Trade expansion				Trade stagnation				Difference in total cost: trade expansion – trade stagnation
	Cost of investment in new capacity	Cost of rehabilitation of existing capacity	Variable cost	Total cost	Cost of investment in new capacity	Cost of rehabilitation of existing capacity	Variable cost	Total cost	
Angola	359	49	78	486	558	50	177	785	−299
Benin	67	19	93	179	74	19	130	223	−44
Botswana	61	14	39	115	246	14	201	461	−346
Burkina Faso	50	21	71	142	50	21	71	142	0
Burundi	80	2	4	87	140	2	6	148	−61
Cameroon	595	52	97	744	328	52	72	452	292
Cape Verde	9	1	15	24	—	—	—	—	24
Central African Republic	48	2	5	56	48	2	5	56	0
Chad	39	0	1	40	72	1	85	157	−117
Congo, Dem. Rep.	1,275	148	49	1,472	526	148	49	723	749
Congo, Rep.	431	7	43	482	559	7	188	754	−272
Côte d'Ivoire	614	81	131	826	644	81	239	964	−138
Equatorial Guinea	3	0	0	3	11	0	1	12	−9
Ethiopia	3,003	102	276	3,381	2,001	102	178	2,281	1,100
Gabon	36	13	13	62	34	17	64	115	−53
Gambia, The	17	3	38	58	17	3	31	51	7
Ghana	560	42	126	728	588	42	624	1,255	−527

Guinea	947	3	98	1,049	161	3	18	183	866
Guinea-Bissau	11	1	9	20	17	1	11	29	-9
Kenya	676	69	274	1,019	697	69	428	1,194	-175
Lesotho	17	2	7	25	17	2	7	25	0
Liberia	39	3	5	48	228	4	18	250	-202
Madagascar	205	6	267	478	—	—	—	—	478
Malawi	34	9	14	56	167	9	14	190	-134
Mali	112	20	47	179	199	21	82	303	-124
Mauritania	29	7	39	74	41	7	97	145	-71
Mauritius	16	9	29	54	—	—	—	—	54
Mozambique	681	30	60	771	465	30	190	685	86
Namibia	108	61	115	284	207	61	315	583	-299
Niger	34	18	24	76	106	20	59	185	-109
Nigeria	4,246	662	2,828	7,736	4,244	659	2,691	7,594	142
South Africa	4,069	1,846	7,596	13,510	4,306	1,846	7,982	14,134	-624
Senegal	369	122	501	993	385	122	555	1,062	-69
Sierra Leone	36	3	28	66	85	3	31	118	-52
Sudan	1,517	43	187	1,748	485	43	135	663	1,085
Tanzania	579	52	280	911	534	52	176	762	149
Togo	97	5	11	113	109	5	105	219	-106
Uganda	498	38	65	601	353	38	56	448	153
Zambia	234	138	99	471	394	138	99	631	-160
Zimbabwe	650	257	302	1,210	577	257	408	1,243	-33
SSA: Total	**22,422**	**3,953**	**13,925**	**40,303**	**19,632**	**3,944**	**15,501**	**39,080**	**1,223**
CAPP: Total	**1,152**	**74**	**159**	**1,387**	**1,052**	**79**	**415**	**1,546**	**-159**
EAPP/Nile									
** Basin: Total**	**6,353**	**306**	**1,086**	**7,747**	**4,210**	**306**	**979**	**5,496**	**2,251**
SAPP: Total	**7,488**	**2,554**	**8,359**	**18,400**	**7,463**	**2,555**	**9,442**	**19,460**	**-1,060**
WAPP: Total	**7,208**	**1,004**	**4,025**	**12,237**	**6,907**	**1,004**	**4,665**	**12,578**	**-341**

Source: Rosnes and Vennemo 2008.
Note: CAPP = Central African Power Pool; EAPP/Nile Basin = East African/Nile Basin Power Pool; SAPP = Southern African Power Pool; SSA = Sub-Saharan Africa; WAPP = West African Power Pool. — = data not available.

Investment Requirements

Table A3.1   Power Demand, Projected Average Annual Growth Rate

	Annual population growth (%)	Base growth scenario, % growth		Low growth scenario, % growth	
		GDP/capita	Electricity demand	GDP/capita	Electricity demand
SAPP					
Angola	2.8	6.3	9.7	3.6	6.4
Botswana	−0.4	4.6	5.2	2.5	2.1
Congo, Dem. Rep.	2.5	1.5	4.2	0.8	3.4
Lesotho	−0.3	4.3	5.0	2.4	2.1
Malawi	2.2	1.7	3.6	0.9	2.5
Mozambique	1.7	1.9	3.3	1.0	2.1
Namibia	1.0	4.0	5.5	2.2	3.0
South Africa	0.1	3.7	4.3	2.0	1.8
Zambia	1.7	2.2	3.6	1.1	2.2
Zimbabwe	0.6	2.7	3.3	1.4	1.5
Weighted average	1.6	2.7	4.5	1.5	2.8
EAPP/Nile Basin					
Burundi	3.2	1.3	4.2	0.7	3.4
Djibouti	1.9	1.4	2.9	0.7	2.0
Egypt, Arab Rep.	1.8	2.3	3.9	1.2	2.5
Ethiopia	2.3	2.6	4.8	1.4	3.2
Kenya	2.6	1.5	3.7	0.8	2.8
Rwanda	2.2	2.7	4.9	0.6	2.2
Sudan	2.0	2.8	4.8	1.5	3.1
Tanzania	1.8	2.3	3.9	1.2	2.4
Uganda	3.8	1.2	4.7	0.6	4.0
Weighted average	2.3	2.2	4.3	1.2	2.9
WAPP					
Benin	2.5	1.8	4.0	0.9	2.9
Burkina Faso	3.0	0.4	2.8	0.2	2.6
Côte d'Ivoire	1.9	1.0	2.3	0.5	1.7
Gambia, The	2.7	1.3	3.5	0.6	2.7
Ghana	1.9	3.2	5.2	1.6	3.2
Guinea	2.6	1.6	3.9	0.8	2.9
Guinea-Bissau	2.0	0.2	1.4	0.1	1.3
Liberia	3.1	1.6	4.3	0.8	3.4
Mali	2.8	1.9	4.5	1.0	3.4
Mauritania	2.9	4.9	8.1	2.5	5.5
Niger	2.9	0.3	2.5	0.1	2.3
Nigeria	2.4	3.2	5.7	1.6	3.8
Senegal	2.5	2.0	4.3	1.0	3.1
Sierra Leone	2.3	0.5	2.1	0.3	1.8
Togo	2.7	0.2	2.2	0.1	2.1

(continued next page)

Table A3.1   *(continued)*

	Annual population growth (%)	Base growth scenario, % growth		Low growth scenario, % growth	
		GDP/capita	Electricity demand	GDP/capita	Electricity demand
Weighted average	2.4	2.4	4.7	1.2	3.3
CAPP					
Cameroon	2.2	2.0	3.9	1.0	2.7
Central African Republic	1.5	1.5	2.6	0.8	1.5
Chad	2.9	0.8	3.2	0.4	2.7
Congo, Rep.	2.8	2.2	4.8	1.1	3.5
Equatorial Guinea	2.0	3.1	5.1	1.6	3.3
Gabon	2.0	1.2	2.7	0.6	1.9
Weighted average	2.4	1.6	3.6	0.8	2.6
Island States					
Cape Verde	0.5	4.7	6.0	2.3	3.1
Madagascar	2.5	1.2	3.2	0.6	2.5
Mauritius	0.8	2.9	3.8	1.5	1.9
Weighted average	2.3	1.4	3.3	0.7	2.5

Source: Rosnes and Vennemo 2008.
Note: CAPP = Central African Power Pool; EAPP/Nile Basin = East African/Nile Basin Power Pool; SAPP = Southern African Power Pool; WAPP = West African Power Pool; GDP = gross domestic product.

Table A3.2   Suppressed Demand for Power

	Outages (hours per year)	Average duration of outages (hours)	Down time (% of a year)	Suppressed demand in 2005 (GWh)
SAPP				
Angola	1,780.8	19.31	20.3	435
Botswana	38.9	1.86	0.4	11
Congo, Dem. Rep.	659.2	3.63	7.5	351
Lesotho	177.9	7.65	2.0	8
Malawi	328.1	4.27	3.7	49
Mozambique	350.4	6.08	4.0	450
Namibia	46.1	2.32	0.5	13
South Africa	24.5	4.15	0.3	602
Zambia	219.9	5.48	2.5	157
Zimbabwe	350.4	6.08	4.0	512

(continued next page)

Table A3.2   *(continued)*

	Outages (hours per year)	Average duration of outages (hours)	Down time (% of a year)	Suppressed demand in 2005 (GWh)
Average for available sample	**350.4**	**6.1**	**4.0**	
EAPP/Nile Basin				
Burundi	1,461.5	10.34	16.7	25
Djibouti	456.4	5.88	5.2	12
Egypt, Arab Rep.	43.3	2.48	0.5	417
Ethiopia	456.4	5.88	5.2	109
Kenya	702.6	8.20	8.0	366
Rwanda	346.9	4.47	4.0	5
Sudan	456.4	5.88	5.2	168
Tanzania	435.9	6.46	5.0	208
Uganda	463.8	6.55	5.3	84
Average for available sample	**456.4**	**5.88**	**5.2**	
WAPP				
Benin	505	2.72	6	34
Burkina Faso	196	1.61	2	11
Côte d'Ivoire	1,101	5.94	13	365
Gambia, The	1,961	6.86	22	29
Ghana	1,465	12.59	17	979
Guinea	2,759	6.78	31	224
Guinea-Bissau	1,978	17.94	23	14
Liberia	1,101	5.94	41	123
Mali	453	2.44	5	21
Mauritania	129	2.89	1	3
Niger	124	0.50	1	6
Nigeria	1,101	5.94	64	10,803
Senegal	1,052	5.67	17	250
Sierra Leone	1,101	5.94	82	189
Togo	1,101	5.94	13	73
Average for available sample	**1,176**	**5.94**	**22**	
CAPP				
Cameroon	613	4.03	7.0	241
Central African Republic	950	5.20	10.8	11

(continued next page)

Table A3.2   *(continued)*

	Outages (hours per year)	*Average duration of outages (hours)*	*Down time (% of a year)*	*Suppressed demand in 2005 (GWh)*
Chad	950	5.20	10.8	10
Congo, Rep.	924	4.33	10.6	616
Equatorial Guinea	950	5.20	10.8	3
Gabon	950	5.20	10.8	134
Average for available sample	**889**	**5.20**	**10.2**	
Island States				
Cape Verde	797.0	5.30	9.0	4
Madagascar	221.1	2.67	2.5	21
Mauritius	1,321.0	7.23	15.1	35

Source: Rosnes and Vennemo 2008.

Note: Market demand for power is one of three categories of demand for power, the others being social demand or access, and suppressed demand. Because of a lack of data, regional sample averages are applied to the following countries: Mozambique, Zimbabwe (SAPP); Ethiopia, Sudan, and Djibouti (EAPP/Nile Basin); Benin, Côte d'Ivoire, Liberia, Mali, Nigeria, Senegal, Sierra Leone, and Togo (WAPP). For CAPP, data are available for Cameroon and Republic of Congo only; for the other countries, regional average is applied. CAPP = Central African Power Pool; EAPP/Nile Basin = East African/Nile Basin Power Pool; SAPP = South African Power Pool; WAPP = West African Power Pool. GWh = gigawatt-hour.

Table A3.3   Target Access to Electricity, by Percentage of Population

% of population

	2005 access			Regional targets			National targets		
	Total	*Urban*	*Rural*	*Total*	*Urban*	*Rural*	*Total*	*Urban*	*Rural*
SAPP									
Angola	14	26	4	24	42	9	46	84	15
Botswana	30	45	9	40	59	14	100	100	100
Congo, Dem. Rep.	8	16	2	20	37	8	39	76	12
Lesotho	6	23	1	17	68	4	35	121	13
Malawi	7	29	1	18	76	1	15	56	3
Mozambique	14	26	2	23	41	5	20	37	5
Namibia	37	75	12	45	86	19	53	95	25
South Africa	71	80	50	79	87	66	100	100	100
Zambia	20	45	3	29	57	11	29	50	15
Zimbabwe	41	87	8	49	99	14	67	100	44
Total SAPP	**26**	**45**	**11**	**37**	**60**	**17**	**51**	**79**	**27**
EAPP/Nile Basin									
Burundi	6	45	0	25	100	13	31	67	25
Djibouti	31	34	5	50	54	11	53	56	29
Egypt, Arab Rep.	93	100	87	100	100	100	100	100	100
Ethiopia	15	76	0	32	100	16	60	100	50
Kenya	27	48	4	42	72	10	67	100	32
Rwanda	16	39	1	29	66	4	18	39	4
Sudan	34	56	12	50	77	23	60	100	21
Tanzania	13	27	1	30	63	2	29	38	22
Uganda	8	44	2	27	100	15	25	100	13
Total EAPP/Nile Basin	**35**	**64**	**19**	**50**	**84**	**31**	**60**	**88**	**45**
WAPP									
Benin	25	50	6	35	68	9	50	100	11

Burkina Faso	9	40	0	20	84	1	23	100	6
Côte d'Ivoire	54	86	23	63	96	30	73	100	46
Gambia, The	59	82	21	66	88	31	79	100	37
Ghana	55	82	21	62	89	30	76	100	37
Guinea	21	54	2	31	77	3	40	100	3
Guinea-Bissau	14	41	2	25	74	4	33	100	6
Liberia	1	0	2	13	1	4	66	100	6
Mali	15	37	2	25	60	5	39	100	7
Mauritania	23	50	3	34	73	4	46	100	6
Niger	7	37	0	18	93	1	20	100	1
Nigeria	59	84	28	67	89	40	82	100	49
Senegal	34	69	6	44	88	9	51	100	10
Sierra Leone	41	82	2	46	91	4	51	100	6
Togo	21	41	2	30	58	5	50	100	8
Total WAPP	**45**	**75**	**16**	**53**	**85**	**23**	**66**	**100**	**34**
CAPP									
Cameroon	61	85	21	68	90	31	71	84	49
Central African Republic	3	8	0	15	36	1	34	84	1
Chad	3	9	0	15	46	1	26	84	0
Congo, Rep.	22	35	0	33	52	0	53	84	0
Equatorial Guinea	11	26	0	22	54	0	35	84	0
Gabon	82	85	54	92	93	83	91	84	82
Total CAPP	**35**	**59**	**8**	**44**	**73**	**12**	**53**	**84**	**19**
Island States									
Cape Verde	55	72	24	62	80	31	84	100	54
Madagascar	18	48	5	36	91	12	36	89	13
Mauritius	100	100	100	100	100	100	100	100	100
Total islands	**23**	**52**	**9**	**40**	**92**	**16**	**40**	**90**	**17**

Source: Rosnes and Vennemo 2008.

Note: CAPP = Central African Power Pool; EAPP/Nile Basin = East African/Nile Basin Power Pool; SAPP = Southern African Power Pool; WAPP = West African Power Pool.

220

Table A3.4 Target Access to Electricity, by Number of New Connections
number of new connections

	2005 access			Regional targets			National targets		
	Total	Urban	Rural	Total	Urban	Rural	Total	Urban	Rural
SAPP									
Angola	189,517	177,270	12,247	626,062	495,006	131,056	1,544,540	1,270,679	273,861
Botswana	4,082	4,082	0	36,618	31,052	5,566	239,160	110,534	128,626
Congo, Dem. Rep.	405,946	387,713	18,234	2,158,523	1,681,848	476,675	4,986,424	4,118,393	868,031
Lesotho	1,834	1,834	0	41,612	34,804	6,808	104,828	73,248	31,580
Malawi	80,533	76,926	3,608	417,587	412,275	5,312	319,224	270,837	48,387
Mozambique	200,355	200,355	0	621,455	546,922	74,533	514,437	445,893	68,544
Namibia	32,212	32,148	64	69,050	51,026	18,024	103,186	67,936	35,250
South Africa	411,428	411,428	0	1,219,077	854,276	364,801	3,189,412	1,612,810	1,576,602
Zambia	126,822	122,068	4,754	386,443	255,967	130,476	380,456	177,410	203,046
Zimbabwe	182,538	182,538	0	396,442	313,325	83,117	896,805	326,552	570,253
Total SAPP	**1,635,267**	**1,596,360**	**38,907**	**5,972,869**	**4,676,501**	**1,296,368**	**12,278,472**	**8,474,292**	**3,804,180**
EAPP/Nile Basin									
Burundi	68,320	66,212	2,108	473,983	231,661	242,322	610,986	132,306	478,680
Djibouti	10,887	10,887	0	46,190	45,142	1,049	52,787	48,058	4,729
Egypt, Arab Rep.	2,498,342	1,478,600	1,019,742	3,746,409	1,478,600	2,267,809	3,746,409	1,478,600	2,267,809
Ethiopia	1,033,168	1,022,756	10,411	4,316,696	1,937,724	2,378,972	9,699,988	1,937,724	7,762,265
Kenya	830,330	818,995	11,335	2,179,612	1,934,176	245,436	4,426,566	3,221,815	1,204,751
Rwanda	201,891	201,891	0	483,356	446,265	37,092	239,596	201,891	37,705
Sudan	798,611	777,563	21,048	2,198,508	1,677,706	520,802	3,110,220	2,681,012	429,208
Tanzania	382,227	381,490	737	1,944,457	1,895,796	48,661	1,839,775	823,674	1,016,101
Uganda	260,351	208,876	51,475	1,845,192	877,476	967,716	1,680,515	877,476	803,039

	6,084,126	4,967,270	1,116,856	17,234,404	10,524,545	6,709,859	25,406,842	11,402,556	14,004,287
Total EAPP/ Nile Basin									
WAPP									
Benin	139,684	130,235	9,449	338,376	290,292	48,084	635,693	571,764	63,929
Burkina Faso	133,569	132,376	1,193	508,094	496,918	11,176	636,793	631,201	5,592
Côte d'Ivoire	493,151	449,496	43,656	871,445	660,116	211,329	1,285,820	742,790	543,030
Gambia, The	72,255	69,774	2,481	102,539	84,114	18,424	155,214	116,041	39,173
Ghana	685,300	669,002	16,298	1,088,276	852,221	236,055	1,792,067	1,182,491	609,576
Guinea	169,952	166,219	3,733	400,901	379,423	21,478	621,287	597,188	24,099
Guinea-Bissau	10,156	9,575	581	49,891	45,852	4,039	77,215	73,428	3,786
Liberia	621	128	493	96,202	6,238	89,964	512,237	507,613	4,624
Mali	157,054	149,790	7,264	458,506	403,796	54,710	886,285	842,148	44,138
Mauritania	53,747	51,306	2,440	142,323	132,492	9,832	240,357	228,690	11,668
Niger	79,566	78,400	1,166	441,127	429,473	11,654	479,434	474,300	5,134
Nigeria	5,228,457	4,930,767	297,690	7,762,673	5,726,037	2,036,636	12,518,720	7,797,915	4,720,806
Senegal	274,605	256,809	17,796	583,312	519,770	63,542	785,043	679,618	105,426
Sierra Leone	205,924	204,913	1,011	277,461	269,994	7,466	345,993	332,753	13,240
Togo	99,647	97,346	2,301	230,519	212,066	18,453	510,621	490,928	19,693
Total WAPP	**7,803,687**	**7,396,135**	**407,552**	**13,351,645**	**10,508,801**	**2,842,844**	**21,482,781**	**15,268,867**	**6,213,913**
CAPP									
Cameroon	680,187	674,275	5,912	977,402	810,180	167,222	1,112,508	658,183	454,325
Central African Republic	6,055	5,881	174	122,722	115,646	7,076	308,692	306,516	2,176
Chad	28,363	28,013	350	327,974	317,986	9,988	616,852	614,860	1,991
Congo, Rep.	57,978	57,911	66	163,163	162,836	326	354,860	354,444	416
Equatorial Guinea	3,021	3,017	4	17,875	17,839	36	33,922	33,893	29
Gabon	55,792	55,791	1	90,583	80,555	10,028	86,592	52,504	34,089
Total CAPP	**831,395**	**824,888**	**6,507**	**1,699,718**	**1,505,042**	**194,675**	**2,513,427**	**2,020,401**	**493,026**

(continued next page)

Table A3.4 (continued)

	2005 access			Regional targets			National targets		
	Total	Urban	Rural	Total	Urban	Rural	Total	Urban	Rural
Island States									
Cape Verde	6,290	6,280	10	12,899	10,596	2,303	31,655	22,142	9,513
Madagascar	249,261	218,924	30,337	1,107,461	855,885	251,576	1,099,554	821,762	277,792
Mauritius	19,567	12,813	6,754	19,567	12,813	6,754	19,567	12,813	6,754
Total islands	**275,118**	**238,017**	**37,101**	**1,139,928**	**879,295**	**260,633**	**1,150,776**	**856,717**	**294,059**
Total Sub-Saharan Africa	**16,629,592**	**15,022,670**	**1,606,922**	**39,398,563**	**28,094,184**	**11,304,379**	**62,832,299**	**38,022,833**	**24,809,466**

Source: Rosnes and Vennemo 2008.
Note: CAPP = Central African Power Pool; EAPP/Nile Basin = East African/Nile Basin Power Pool; SAPP = Southern African Power Pool; WAPP = West African Power Pool.

Table A3.5   Total Electricity Demand
TWh

	Total net demand in 2005	Market demand 2015[a]	Social demand with national targets 2015	Total net demand 2015	Increase in net demand 2005–15 (%)
SAPP					
Angola	2.1	6.0	1.9	7.9	375
Botswana	2.4	4.0	0.2	4.2	174
Congo, Dem. Rep.	4.7	7.4	6.2	13.6	288
Lesotho	0.4	0.8	0.1	0.9	224
Malawi	1.3	1.9	0.4	2.3	176
Mozambique	11.2	15.7	0.7	16.4	145
Namibia	2.6	4.2	0.1	4.3	164
South Africa	215.0	316.0	3.2	319.2	147
Zambia	6.3	9.0	0.4	9.3	147
Zimbabwe	12.8	18.0	0.8	18.7	145
Total	**258.8**	**383.0**	**14.0**	**396.9**	152
EAPP/Nile Basin					
Burundi	0.2	0.3	0.5	0.7	349
Djibouti	0.2	0.3	0.1	0.4	199
Egypt, Arab Rep.	84.4	119.9	3.4	123.3	145
Ethiopia	2.1	3.4	7.4	10.7	509
Kenya	4.6	6.8	5.2	12.0	260
Rwanda	0.1	0.2	0.3	0.5	499
Sudan	3.2	5.2	3.9	9.2	287
Tanzania	4.2	6.2	1.7	7.9	187
Uganda	1.6	2.5	1.7	4.2	262
Total	**100.6**	**144.8**	**24.2**	**169.0**	167
WAPP					
Benin	0.6	0.9	0.8	1.7	282
Burkina Faso	0.5	0.6	0.9	1.5	299
Côte d'Ivoire	2.9	4.0	1.4	5.4	185
Gambia, The	0.1	0.2	0.2	0.4	399
Ghana	5.9	10.8	2.0	12.8	216
Guinea	0.7	1.3	0.8	2.2	313
Guinea-Bissau	0.1	0.1	0.1	0.2	199
Liberia	0.3	0.6	0.7	1.3	432
Mali	0.4	0.6	1.2	1.8	449
Mauritania	0.2	0.5	0.3	0.8	399
Niger	0.4	0.6	0.7	1.2	299
Nigeria	16.9	45.6	13.6	59.2	349
Senegal	1.5	2.5	1.0	3.5	232
Sierra Leone	0.2	0.5	0.5	1.0	499
Togo	0.6	0.8	0.7	1.5	249
Total	**31.3**	**69.6**	**24.8**	**94.3**	300
CAPP					
Cameroon	3.4	5.2	1.2	6.4	187
Central African Republic	0.1	0.1	0.4	0.6	599
Chad	0.1	0.1	0.8	1.0	999
Congo, Rep.	5.8	9.8	0.5	10.3	175
Equatorial Guinea	0.03	0.1	0.05	0.1	332
Gabon	1.2	1.7	0.1	1.8	149
Total	**10.7**	**17.1**	**3.1**	**20.2**	188

(continued next page)

Table A3.5 *(continued)*

	Total net demand in 2005	Market demand 2015[a]	Social demand with national targets 2015	Total net demand 2015	Increase in net demand 2005–15 (%)
Island States					
Cape Verde	0.04	0.1	0.04	0.1	249
Madagascar	0.8	1.1	1.4	2.6	324
Mauritius	0.2	0.4	0.02	0.4	199
Total	**1.04**	**1.6**	**1.46**	**3.0**	272

Source: Rosnes and Vennemo 2008.
Note: CAPP = Central African Power Pool; EAPP/Nile Basin = East African/Nile Basin Power Pool; SAPP = Southern African Power Pool; WAPP = West African Power Pool.
a. Assuming all suppressed demand is met.

Table A3.6   Generating Capacity in 2015 under Various Trade, Access, and Growth Scenarios

Generation capacity (MW)	Trade expansion scenario			Trade stagnation scenario	Low-growth scenario
	Constant access rate	Regional target access rate	National targets for access rates	National targets for access rates	National targets for access rates, trade expansion
SAPP					
Installed capacity[a]	17,136	17,136	17,136	17,136	17,136
Refurbished capacity	28,029	28,035	28,046	28,148	28,046
New capacity	31,297	32,168	33,319	32,013	20,729
Hydropower share (%)	33	33	34	25	40
EAPP/Nile Basin					
Installed capacity[a]	22,132	22,132	22,132	22,132	22,132
Refurbished capacity	1,369	1,375	1,375	1,381	1,375
New capacity	23,045	24,639	25,637	17,972	23,540
Hydropower share (%)	49	47	48	28	48
WAPP					
Installed capacity[a]	4,096	4,096	4,096	4,096	4,096
Refurbished capacity	5,530	6,162	6,972	6,842	5,535
New capacity	15,979	16,634	18,003	16,239	17,186
Hydropower share (%)	82	79	77	73	80

(continued next page)

Table A3.6   (continued)

| Generation capacity (MW) | Trade expansion scenario | | | Trade stagnation scenario | Low-growth scenario |
	Constant access rate	Regional target access rate	National targets for access rates	National targets for access rates	National targets for access rates, trade expansion
CAPP					
Installed capacity[a]	260	260	260	260	260
Refurbished capacity	906	906	906	1,081	906
New capacity	3,856	4,143	4,395	3,833	3,915
Hydropower share (%)	97	97	97	83	97
Island States					
Installed capacity[a]	282	282	282	282	282
Refurbished capacity	83	83	83	83	83
New capacity	189	369	368	368	353
Hydropower share (%)	25	19	19		20
Total Sub-Saharan Africa					
Installed capacity[a]	43,906	43,906	43,906	43,906	43,906
Refurbished capacity	35,917	36,561	37,382	37,535	35,945
New capacity	74,366	77,953	81,722	70,425	65,723
Hydropower share (%)	48	47	47	36	52

Source: Rosnes and Vennemo 2008.
Note: CAPP = Central African Power Pool; EAPP/Nile Basin = East African/Nile Basin Power Pool; SAPP = Southern African Power Pool; WAPP = West African Power Pool.
a. "Installed capacity" refers to installed capacity as of 2005 that will not undergo refurbishment before 2015. Existing capacity that will be refurbished before 2015 is not included in the installed capacity figure, but in the refurbishment figure.

Table A3.7a Annualized Costs of Capacity Expansion, Constant Access Rates, Trade Expansion
$ million

	Generation			T&D						Total			Total costs of capacity expansion
	Invest-ment	Rehabil-itation	Variable	Cross-border grid	Distribu-tion grid	Urban connec-tion	Rural connec-tion	Rehab-ilitation	Variable	Invest-ment	Rehabili-tation	Variable	
Angola	0.1	10.5	0	78	139	11	2	38	78	231	49	78	357
Benin	0	4	57	0	18	8	2	15	36	28	19	93	140
Botswana	0	0.6	10.3	12	19	0	0	14	29	30	14	39	84
Burkina Faso	0	4	24	3	10	8	0	17	32	21	21	57	99
Burundi	0	1	0.4	0	3	4	0	1	4	7	2	4	14
Cameroon	410	32	54	22	22	42	1	21	43	497	52	97	646
Cape Verde	3	0	10	0	1	0	0	1	1	4	1	11	16
Central African Republic	5	1	1	0	1	0	0	1	2	6	2	3	12
Chad	0	0	0	0	0	2	0	0	1	2	0	1	4
Congo, Dem. Rep.	652	110	0	53	29	24	3	38	49	761	148	49	958
Congo, Rep.	301	7	24	20	9	4	0	0	13	332	7	37	376
Côte d'Ivoire	40	32	12	32	38	28	8	49	99	146	81	111	338
Equatorial Guinea	0	0	0	0	0	0	0	0	0	1	0	0	1
Ethiopia	1,136	18	133	71	69	64	2	84	116	1342	102	249	1,693
Gabon	19	9	5	2	3	3	0	5	8	27	13	13	54
Gambia, The	0	2	30	0	1	4	0	1	1	6	3	31	40
Ghana	262	9	32	5	79	42	3	34	94	391	42	126	559
Guinea	860	1	91	44	6	10	1	2	7	920	3	98	1,022
Guinea-Bissau	0	1	8	2	0	1	0	0	1	3	1	8	12
Kenya	231	20	144	15	65	51	1	48	130	362	68	274	705
Lesotho	0	0	0	0	6	0	0	2	7	6	2	7	15
Liberia	0	3	1	1	5	0	0	0	4	5	3	5	14

Madagascar	33	4	142	0	2	14	9	2	6	58	6	148	212
Malawi	0	3.8	0	1	7	5	1	5	14	13	9	14	35
Mali	0	2	4	35	24	9	1	18	43	70	20	47	137
Mauritania	0	2	29	0	11	3	0	5	11	15	7	39	61
Mauritius	0	0	8	0	13	1	2	9	21	16	9	29	54
Mozambique	606	9.5	17.4	17	17	12	0	20	43	653	30	60	742
Namibia	0	6.4	13.1	29	67	2	0	55	102	98	61	115	274
Niger	0	2	10	1	4	5	0	16	14	10	18	24	52
Nigeria	1,671	169	1,452	1	1,013	306	55	473	596	3,046	642	2,048	5,736
Senegal	41	4	131	1	261	16	3	118	370	322	122	501	946
Sierra Leone	0	0	6	3	5	13	0	1	5	21	1	10	33
South Africa	2,607	1,046	5,026	2	1,173	18	—	800	2,495	3,800	1,846	7,521	13,167
Sudan	1,166	—	88	50	59	48	4	43	97	1327	43	186	1,556
Tanzania	243	16	132	10	29	24	0	36	60	306	52	192	550
Togo	0	2	0	0	3	6	0	3	7	10	5	8	22
Uganda	289	10	27	16	23	13	6	28	39	347	38	65	451
Zambia	0	57	0	141	43	8	1	81	99	193	138	99	429
Zimbabwe	403	97	117	41	75	0	0	160	185	518	257	302	1,077
SSA: Total	**10,978**	**1,696**	**7,839**	**708**	**3,352**	**809**	**105**	**2,244**	**4,962**	**15,951**	**3,937**	**12,799**	**32,693**
CAPP: Total	**1,901**	**61**	**172**	**172**	**236**	**114**	**7**	**109**	**246**	**2,430**	**168**	**419**	**3,020**
EAPP/Nile Basin: Total	**2,852**	**182**	**460**	**185**	**227**	**184**	**12**	**235**	**411**	**3,457**	**417**	**870**	**4,747**
SAPP: Total	**4,544**	**1,361**	**5,458**	**384**	**1,606**	**118**	**16**	**1,251**	**3,167**	**6,667**	**2,612**	**8,624**	**17,900**
WAPP: Total	**2,877**	**235**	**1,868**	**128**	**1,468**	**456**	**73**	**748**	**1,310**	**5,003**	**982**	**3,178**	**9,166**

Source: Rosnes and Vennemo 2008.

Note: CAPP = Central African Power Pool; EAPP/Nile Basin = East African/Nile Basin Power Pool; SAPP = Southern African Power Pool; SSA = Sub-Saharan Africa; WAPP = West African Power Pool; T&D = transmission and distribution. — = Not available.

Table A3.7b Annualized Costs of Capacity Expansion, 35% Access Rates, Trade Expansion
$ million

	Generation			T&D						Total			Total costs of capacity expansion
	Investment	Rehabilitation	Variable	Cross-border grid	Distribution grid	Urban connection	Rural connection	Rehabilitation	Variable	Investment	Rehabilitation	Variable	
Angola	0.7	10.5	0	81	139	31	24	38	78	277	49	78	403
Benin	1	4	57	0	18	18	9	15	36	46	19	93	158
Botswana	0.1	0.6	10.3	10	19	2	1	14	29	31	14	39	85
Burkina Faso	0	4	39	1	10	31	2	17	32	44	21	71	136
Burundi	5.9	1	0.4	1	3	14	29	1	4	53	2	4	59
Cameroon	415	32	54	21	22	50	31	21	43	539	52	97	688
Cape Verde	3	0	11	0	1	1	1	1	1	5	1	12	18
Central African Republic	13	1	1	0	1	7	1	1	2	23	2	4	29
Chad	0	0	0	0	0	20	2	0	1	22	0	1	23
Congo, Dem. Rep.	697	110	0	53	29	104	89	38	49	972	148	49	1,169
Congo, Rep.	351	7	28	17	9	10	0	0	13	387	7	41	434
Côte d'Ivoire	102	32	15	28	38	41	39	49	99	249	81	114	445
Equatorial Guinea	0	0	0	0	0	1	0	0	0	2	0	0	2
Ethiopia	1,200	18	141	66	69	120	431	84	116	1,887	102	257	2,246
Gabon	20	9	5	2	3	5	2	5	8	32	13	13	59
Gambia, The	1	2	36	0	1	5	3	1	1	10	3	37	50
Ghana	272	9	32	4	79	53	44	34	94	452	42	126	620
Guinea	860	1	91	43	6	24	4	2	7	936	3	98	1,038
Guinea-Bissau	0	1	8	3	0	3	1	0	1	7	1	9	17
Kenya	236	21	144	0	65	120	29	48	130	450	69	274	792

Lesotho	0	0	0	0	6	2	1	2	7	10	2	7	18
Liberia	4	3	1	1	5	0	17	0	4	26	3	5	35
Madagascar	70	4	266	0	2	53	74	2	6	200	6	272	477
Malawi	0	3.8	0	1	7	26	1	5	14	34	9	14	57
Mali	2	2	4	32	24	25	10	18	43	94	20	47	161
Mauritania	0	2	29	1	11	8	2	5	11	22	7	39	68
Mauritius	0	0	8	0	13	1	2	9	21	16	9	29	54
Mozambique	606	9.5	17.4	17	17	34	14	20	43	688	30	60	778
Namibia	0.4	6.4	13.1	29	67	3	3	55	102	103	61	115	279
Niger	0	2	10	1	4	27	2	16	14	33	18	24	75
Nigeria	1,748	177	1,779	1	1,013	355	380	473	596	3,497	651	2,376	6,523
Senegal	42	4	131	2	261	32	12	118	370	349	122	501	972
Sierra Leone	0	1	23	3	5	17	1	1	5	27	3	28	57
South Africa	2,677	1,046	5,072	1	1,173	37	18	800	2,495	3,907	1,846	7,567	13,319
Sudan	1,179	0	90	39	59	104	94	43	97	1,476	43	187	1707
Tanzania	255	16	181	0	29	118	9	36	60	411	52	241	704
Togo	60	2	4	0	3	13	3	3	7	80	5	11	95
Uganda	317	10.1	26.6	10	23	54	116	28	39	520	38	65	624
Zambia	0.6	56.5	0	142	43	16	24	81	99	226	138	99	463
Zimbabwe	404	97	117	36	75	19	15	160	185	549	257	302	1,109
SSA: Total	**11,543**	**1,705**	**8,445**	**646**	**3,352**	**1,604**	**1,540**	**2,244**	**4,962**	**18,692**	**3,949**	**13,406**	**36,046**
CAPP: Total	**1,985**	**61**	**178**	**161**	**236**	**242**	**183**	**109**	**246**	**2,811**	**168**	**425**	**3,404**
EAPP/Nile Basin: Total	**3,062**	**183**	**521**	**147**	**227**	**540**	**703**	**235**	**411**	**4,680**	**418**	**931**	**6,028**
SAPP: Total	**4,711**	**1,360**	**5,677**	**370**	**1,606**	**445**	**273**	**1,251**	**3,167**	**7,408**	**2,612**	**8,843**	**18,861**
WAPP: Total	**3,095**	**244**	**2,241**	**119**	**1,468**	**645**	**528**	**748**	**1,310**	**5,855**	**993**	**3,552**	**10,400**

Source: Rosnes and Vennemo 2008.
Note: CAPP = Central African Power Pool; EAPP/Nile Basin = East African/Nile Basin Power Pool; SAPP = Southern African Power Pool; SSA = Sub-Saharan Africa; WAPP = West African Power Pool; T&D = transmission and distribution.

Table A3.7c Annualized Costs of Capacity Expansion, National Targets for Access Rates, Trade Expansion
$ million

	Generation			T&D						Total			
	Invest-ment	Rehabil-itation	Variable	Cross-border grid	Distribu-tion grid	Urban connec-tion	Rural connec-tion	Rehab-ilitation	Variable	Invest-ment	Rehabili-tation	Variable	Total costs of capacity expansion
Angola	3	11	0	87	139	79	51	38	78	359	49	78	486
Benin	1	4	57	0	18	35	12	15	36	67	19	93	179
Botswana	2	1	10	10	19	7	24	14	29	61	14	39	115
Burkina Faso	0	4	39	0	10	39	1	17	32	50	21	71	142
Burundi	11	1	1	1	3	8	57	1	4	80	2	4	87
Cameroon	426	32	54	21	22	41	85	21	43	595	52	97	744
Cape Verde	4	0	13	0	1	1	3	1	1	9	1	15	24
Central African Republic	28	1	3	0	1	19	0	1	2	48	2	5	56
Chad	0	0	0	0	0	38	0	0	1	39	0	1	40
Congo, Dem. Rep.	775	110	0	54	29	256	162	38	49	1,275	148	49	1,472
Congo, Rep.	385	7	31	15	9	22	0	0	13	431	7	43	482
Côte d'Ivoire	402	32	31	27	38	46	101	49	99	614	81	131	826
Equatorial Guinea	0	0	0	0	0	2	0	0	0	3	0	0	3
Ethiopia	1,345	18	160	61	69	120	1,408	84	116	3,003	102	276	3,381
Gabon	21	9	5	2	3	3	6	5	8	36	13	13	62
Gambia, The	2	2	37	0	1	7	7	1	1	17	3	38	58
Ghana	288	9	32	5	79	73	114	34	94	560	42	126	728
Guinea	860	1	91	40	6	37	4	2	7	947	3	98	1,049
Guinea-Bissau	0	1	8	5	0	5	1	0	1	11	1	9	20
Kenya	264	21	144	3	65	200	144	48	130	676	69	274	1,019
Lesotho	0.1	0	0	0	6	5	6	2	7	17	2	7	25

Liberia	0	3	1	2	5	32	1	0	4	39	3	5	48
Madagascar	70	4	261	0	2	51	82	2	6	205	6	267	478
Malawi	0.3	4	0	1	7	17	9	5	14	34	9	14	56
Mali	1	2	4	26	24	52	8	18	43	112	20	47	179
Mauritania	0	2	29	1	11	14	2	5	11	29	7	39	74
Mauritius	0	0	8	0	13	1	2	9	21	16	9	29	54
Mozambique	606	10	17	17	17	28	13	20	43	681	30	60	771
Namibia	1	6	13	30	67	4	7	55	102	108	61	115	284
Niger	0	2	10	0	4	29	1	16	14	34	18	24	76
Nigeria	1,867	189	2,232	2	1,013	484	880	473	596	4,246	662	2,828	7,736
Senegal	44	4	131	3	261	42	20	118	370	369	122	501	993
Sierra Leone	0	1	23	7	5	21	2	1	5	36	3	28	66
South Africa	2,745	1,046	5,101	1	1,173	70	80	800	2,495	4,069	1,846	7,596	13,510
Sudan	1,177	0	90	37	59	166	78	43	97	1,517	43	187	1,748
Tanzania	303	16	220	12	29	51	184	36	60	579	52	280	911
Togo	60	2	4	0	3	30	4	3	7	97	5	11	113
Uganda	312	10	27	12	23	54	96	28	39	498	38	65	601
Zambia	1	57	0	142	43	11	38	81	99	234	138	99	471
Zimbabwe	412	97	117	37	75	20	106	160	185	650	257	302	1,210
SSA: Total	**12,416**	**1,719**	**9,004**	**661**	**3,352**	**2,220**	**3,799**	**2,244**	**4,962**	**22,451**	**3,960**	**13,964**	**40,377**
CAPP: Total	**2,051**	**61**	**184**	**163**	**236**	**378**	**277**	**109**	**246**	**3,108**	**168**	**428**	**3,708**
EAPP/Nile													
Basin: Total	**3,395**	**183**	**583**	**158**	**227**	**711**	**2,051**	**235**	**411**	**6,542**	**418**	**991**	**7,953**
SAPP: Total	**4,918**	**1,362**	**5,739**	**391**	**1,606**	**599**	**762**	**1,251**	**3,167**	**8,272**	**2,612**	**8,906**	**19,789**
WAPP: Total	**3,529**	**256**	**2,713**	**117**	**1,468**	**933**	**1,159**	**748**	**1,310**	**7,208**	**1,004**	**4,025**	**12,237**

Source: Rosnes and Vennemo 2008.

Note: CAPP = Central African Power Pool; EAPP/Nile Basin = East African/Nile Basin Power Pool; SAPP = Southern African Power Pool; SSA = Sub-Saharan Africa; WAPP = West African Power Pool; T&D = transmission and distribution.

Table A3.7d Annualized Costs of Capacity Expansion, Low Growth Scenario, National Targets for Access Rates, Trade Expansion

$ million

	Generation			T&D						Total			
	Investment	Rehabilitation	Variable	Cross-border grid	Distribution grid	Urban connection	Rural connection	Rehabilitation	Variable	Investment	Rehabilitation	Variable	Total costs of capacity expansion
Angola	2.6	10.8	0	73	87	79	51	38	78	293	49	78	420
Benin	1	4	57	0	13	35	12	15	36	62	19	93	174
Botswana	2.1	0.6	10.3	17	8	7	24	14	29	57	14	39	111
Burkina Faso	0	4	39	0	9	39	1	17	32	50	21	71	142
Burundi	11.4	1	0.5	0	2	8	57	1	4	80	2	4	86
Cameroon	426	32	54	24	15	41	85	21	43	591	52	97	740
Cape Verde	3	0	11	0	1	1	3	1	1	8	1	12	21
Central African Republic	27	1	3	0	1	19	0	1	2	47	2	5	55
Chad	0	0	0	0	0	38	0	0	1	39	0	1	40
Congo, Dem. Rep.	735	110	0.1	53	23	256	162	38	49	1,228	148	49	1,426
Congo, Rep.	278	7	23	18	6	22	0	0	13	325	7	36	367
Côte d'Ivoire	166	32	18	24	30	46	101	49	99	368	81	117	566
Equatorial Guinea	0	0	0	0	0	2	0	0	0	2	0	0	3
Ethiopia	1,345	18	159.5	62	46	120	1,408	84	116	2,981	102	276	3,359
Gabon	19	9	5	2	2	3	6	5	8	33	13	13	59
Gambia, The	2	2	37	0	1	7	7	1	1	17	3	38	57
Ghana	288	9	32	4	52	73	114	34	94	531	42	126	699
Guinea	860	1	91	41	5	37	4	2	7	947	3	98	1,049
Guinea-Bissau	0	1	8	4	0	5	1	0	1	9	1	9	19
Kenya	264	21	144	2	50	200	144	48	130	661	69	274	1,004

													Total
Lesotho	0.1	0	0	0	3	5	6	2	7	14	2	7	22
Liberia	0	3	1	2	2	32	1	0	4	36	3	5	45
Madagascar	68	4	253	0	2	51	82	2	6	202	6	258	466
Malawi	0.3	3.8	0	1	5	17	9	5	14	32	9	14	54
Mali	1	2	4	28	18	52	8	18	43	107	20	47	175
Mauritania	0	2	29	1	7	14	2	5	11	24	7	39	70
Mauritius	0	0	6	0	8	1	2	9	21	11	9	28	47
Mozambique	606	9.5	17.4	16	11	28	13	20	43	673	30	60	763
Namibia	0.7	6.4	13.1	26	34	4	7	55	102	72	61	115	248
Niger	0	2	10	0	4	29	1	16	14	34	18	24	76
Nigeria	1,867	168	1,429	1	343	484	880	473	596	3,576	641	2,026	6,243
Senegal	44	4	131	2	174	42	20	118	370	281	122	501	904
Sierra Leone	0	1	23	6	1	21	2	1	5	31	3	28	61
South Africa	777	1,046	3,820	4	455	70	80	800	2,495	1,385	1,846	6,315	9,546
Sudan	1,177	0	90	37	37	166	78	43	97	1,495	43	187	1,725
Tanzania	285	16	134	9	18	51	184	36	60	548	52	195	794
Togo	60	2	4	0	3	30	4	3	7	97	5	11	113
Uganda	312	10	27	9	19	54	96	28	39	491	38	65	594
Zambia	1	57	0	141	27	11	38	81	99	218	138	99	455
Zimbabwe	412	97	117	52	36	20	106	160	185	626	257	302	1,186
SSA: Total	**10,041**	**1,697**	**6,801**	**659**	**1,558**	**2,220**	**3,799**	**2,244**	**4,962**	**18,282**	**3,939**	**11,762**	**33,984**
CAPP: Total	**1,941**	**61**	**176**	**154**	**150**	**378**	**277**	**109**	**246**	**2,905**	**168**	**421**	**3,495**
EAPP/Nile													
Basin: Total	**3,230**	**183**	**488**	**153**	**164**	**711**	**2,051**	**235**	**411**	**6,314**	**418**	**899**	**7,630**
SAPP: Total	**2,890**	**1,361**	**4,365**	**392**	**709**	**599**	**762**	**1,251**	**3,167**	**5,348**	**2,612**	**7,531**	**15,491**
WAPP: Total	**3,292**	**235**	**1,895**	**112**	**656**	**933**	**1,159**	**748**	**1,310**	**6,154**	**983**	**3,206**	**10,344**

Source: Rosnes and Vennemo 2008.

Note: CAPP = Central African Power Pool; EAPP/Nile Basin = East African/Nile Basin Power Pool; SAPP = Southern African Power Pool; SSA = Sub-Saharan Africa; WAPP = West African Power Pool; T&D = transmission and distribution.

Table A3.7e Annualized Costs of Capacity Expansion, Trade Stagnation

$ million

	Generation			T&D						Total			Total costs of capacity expansion
	Investment	Rehabilitation	Variable	Cross-border grid	Distribution grid	Urban connection	Rural connection	Rehabilitation	Variable	Investment	Rehabilitation	Variable	
Angola	289	12	99	0	139	79	51	38	78	558	50	177	785
Benin	8	4	94	0	18	35	12	15	36	74	19	130	223
Botswana	196	1	172	0	19	7	24	14	29	246	14	201	461
Burkina Faso	0	4	39	0	10	39	1	17	32	50	21	71	142
Burundi	72	1	2	0	3	8	57	1	4	140	2	6	148
Cameroon	180	32	29	0	22	41	85	21	43	328	52	72	452
Cape Verde	—	—	—	—	—	—	—	—	—	—	—	—	—
Central African Republic	28	1	3	0	1	19	0	1	2	48	2	5	56
Chad	33	1	84	0	0	38	0	0	1	72	1	85	157
Congo, Rep.	528	7	175	0	9	22	0	0	13	559	7	188	754
Côte d'Ivoire	458	32	140	0	38	46	101	49	99	644	81	239	964
Congo, Dem. Rep.	80	110	0.1	0	29	256	162	38	49	526	148	49	723
Equatorial Guinea	8	0	0	0	0	2	0	0	0	11	0	1	12
Ethiopia	403	18	62	0	69	120	1,408	84	116	2,001	102	178	2,281
Gabon	21	12	56	0	3	3	6	5	8	34	17	64	115
Gambia, The	2	2	30	0	1	7	7	1	1	17	3	31	51
Ghana	322	9	530	0	79	73	114	34	94	588	42	624	1,255
Guinea	114	1	11	0	6	37	4	2	7	161	3	18	183
Guinea-Bissau	11	1	11	0	0	5	1	0	1	17	1	11	29
Kenya	288	21	298	0	65	200	144	48	130	697	69	428	1,194

Lesotho	0.1	0	0	0	6	5	6	2	7	17	2	7	25
Liberia	191	4	14	0	5	32	1	0	4	228	4	18	250
Madagascar	—	—	—	—	—	—	—	—	—	—	—	—	—
Malawi	135	4	0	0	7	17	9	5	14	167	9	14	190
Mali	115	3	40	0	24	52	8	18	43	199	21	82	303
Mauritania	13	2	87	0	11	14	2	5	11	41	7	97	145
Mauritius	—	—	—	—	—	—	—	—	—	—	—	—	—
Mozambique	407	10	147	0	17	28	13	20	43	465	30	190	685
Namibia	130	6	213	0	67	4	7	55	102	207	61	315	583
Niger	72	4	44	0	4	29	1	16	14	106	20	59	185
Nigeria	1,867	186	2,095	0	1,013	484	880	473	596	4,244	659	2,691	7,594
Senegal	62	4	185	0	261	42	20	118	370	385	122	555	1,062
Sierra Leone	56	1	26	0	5	21	2	1	5	85	3	31	118
South Africa	2,984	1,046	5,487	0	1,173	70	80	800	2,495	4,306	1,846	7,982	14,134
Sudan	182	0	38	0	59	166	78	43	97	485	43	135	663
Tanzania	270	16	115	0	29	51	184	36	60	534	52	176	762
Togo	72	3	97	0	3	30	4	3	7	109	5	105	219
Uganda	179	10	18	0	23	54	96	28	39	353	38	56	448
Zambia	302	57	0	0	43	11	38	81	99	394	138	99	631
Zimbabwe	376	97	223	0	75	20	106	160	185	577	257	408	1,243
SSA: Total	**10,454**	**1,722**	**10,664**	**0**	**3,336**	**2,167**	**3,712**	**2,232**	**4,934**	**19,673**	**3,951**	**15,598**	**39,225**
CAPP: Total	**1,271**	**91**	**451**	**0**	**265**	**402**	**378**	**158**	**332**	**2,320**	**248**	**784**	**3,352**
EAPP/Nile													
Basin: Total	**2,198**	**105**	**810**	**0**	**236**	**501**	**1,990**	**246**	**461**	**4,928**	**351**	**1,271**	**6,551**
SAPP: Total	**5,617**	**1,256**	**6,631**	**0**	**1,584**	**314**	**518**	**1,211**	**3,125**	**8,030**	**2,466**	**9,757**	**20,253**
WAPP: Total	**2,972**	**336**	**3,216**	**0**	**1,458**	**1,142**	**1,217**	**736**	**1,259**	**6,789**	**1,071**	**4,475**	**12,337**

Source: Rosnes and Vennemo 2008.

Note: CAPP = Central African Power Pool; EAPP/Nile Basin = East African/Nile Basin Power Pool; SAPP = Southern African Power Pool; SSA = Sub-Saharan Africa; WAPP = West African Power Pool; T&D = transmission and distribution. — = data not available.

Table A3.8   Annualized Costs of Capacity Expansion under Different Access Rate Scenarios, Trade Expansion
percent of 2005 GDP

| | Trade expansion | | | | Trade stagnation |
| | *Access rate scenarios* | | | | *National targets* |
	Sustain current levels	*Uniform 35% target*	*National targets*	*Low growth, national targets*	
Angola	1.2	1.3	1.6	1.4	2.6
Benin	3.3	3.7	4.2	4.1	5.2
Botswana	0.8	0.8	1.1	1.1	4.4
Burkina Faso	1.8	2.5	2.6	2.6	2.6
Burundi	1.8	7.4	10.9	10.8	18.6
Cameroon	3.9	4.1	4.5	4.5	2.7
Cape Verde	1.6	1.8	2.4	2.1	—
Central African Republic	0.9	2.1	4.1	4.1	4.1
Chad	0.1	0.4	0.7	0.7	2.7
Congo, Rep.	6.2	7.1	7.9	6.0	12.4
Côte d'Ivoire	2.1	2.7	5.1	3.5	5.9
Congo, Dem. Rep.	13.5	16.5	20.7	20.1	10.2
Equatorial Guinea	0.01	0.03	0.04	0.04	0.2
Ethiopia	13.8	18.3	27.5	27.3	18.5
Gabon	0.6	0.7	0.7	0.7	1.3
Gambia, The	8.7	10.8	12.6	12.4	11.1
Ghana	5.2	5.8	6.8	6.5	11.7
Guinea	31.3	31.8	32.2	32.2	5.6
Guinea-Bissau	4.0	5.6	6.6	6.3	9.6
Kenya	3.8	4.2	5.4	5.4	6.4
Lesotho	1.1	1.3	1.8	1.5	1.8
Liberia	2.6	6.6	9.1	8.5	47.2
Madagascar	4.2	9.5	9.5	9.2	—
Malawi	1.2	2.0	2.0	1.9	6.7
Mali	2.6	3.0	3.4	3.3	5.7
Mauritania	3.3	3.7	4.0	3.8	7.9
Mauritius	0.9	0.9	0.9	0.7	—
Mozambique	11.3	11.8	11.7	11.6	10.4
Namibia	4.4	4.5	4.6	4.0	9.4
Niger	1.6	2.3	2.3	2.3	5.6
Nigeria	5.1	5.8	6.9	5.6	6.8
Senegal	10.9	11.2	11.4	10.4	12.2
Sierra Leone	2.7	4.7	5.4	5.0	9.7
South Africa	5.4	5.5	5.6	3.9	5.8

(continued next page)

Table A3.8   *(continued)*

| | Trade expansion | | | Trade stagnation |
| | Access rate scenarios | | | National targets |
	Sustain current levels	*Uniform 35% target*	*National targets*	*Low growth, national targets*	
Sudan	5.7	6.2	6.4	6.3	2.4
Tanzania	3.9	5.0	6.4	5.6	5.4
Togo	1.0	4.4	5.2	5.2	10.2
Uganda	5.2	7.1	6.9	6.8	5.1
Zambia	5.8	6.3	6.4	6.2	8.6
Zimbabwe	31.5	32.4	35.4	34.7	36.4
SSA: Total	**215**	**262**	**303**	**288**	**333**
CAPP: Total	**16**	**25**	**34**	**32**	**41**
EAPP/Nile Basin: Total	**37**	**52**	**70**	**65**	**72**
SAPP: Total	**77**	**88**	**94**	**87**	**104**
WAPP: Total	**96**	**117**	**132**	**127**	**153**

Source: Rosnes and Vennemo 2008.
Note: CAPP = Central African Power Pool; EAPP/Nile Basin = East African/Nile Basin Power Pool; SAPP = Southern African Power Pool; SSA = Sub-Saharan Africa; WAPP = West African Power Pool. — = data not available.

APPENDIX 4

Strengthening Sector Reform and Planning

Table A4.1 Institutional Indicators: Summary Scores by Group of Indicators
Out of 100, 2007

	Reform	Reform sector specific	Regulation	Regulation sector specific	SOE governance	Aggregate score
Benin	13	20	0	28	47	22
Burkina Faso	50	83	50	19	75	55
Cameroon	80	67	48	75	80	70
Cape Verde	69	42	82	31	61	57
Chad	38	42	0	28	60	34
Congo, Dem. Rep.	33	42	0	83	49	41
Cote d'Ivoire	88	83	36	78	70	71
Ethiopia	64	42	56	31	100	59
Ghana	88	58	83	50	70	70
Kenya	86	75	53	78	64	71
Lesotho	79	0	64	0	50	39
Madagascar	67	42	59	67	86	64
Malawi	64	42	61	67	79	63
Mozambique	46	42	0	44	100	46
Namibia	79	42	76	67	73.5	68
Niger	49	50	48	61	74	56
Nigeria	78	58	53	44	56	58

(continued next page)

Table A4.1   *(continued)*

	Reform	Reform sector specific	Regulation	Regulation sector specific	SOE governance	Aggregate score
Rwanda	62	17	53	56	45	47
Senegal	71	58	58	78	61	65
South Africa	88	42	82	44	65.5	64
Sudan	33	0	—	—	50	28
Tanzania	69	58	67	33	64	58
Uganda	84	75	64	33	63	64
Zambia	71	58	64	67	54	63

Source: Vagliasindi and Nellis 2010.
Note: SOE = state-owned enterprise; — = data not available.

Table A4.2a   Institutional Indicators: Description of Reform Indicators

Subindex	Indicator	Indicator values
Legislation	Existence of reform	0 = No reform of the sector 1 = At least one key reform of the sector
	Legal reform	0 = No new sector legislation passed within the past 10 years 1 = New sector legislation passed in the past 10 years
Restructuring	Unbundling	0 = Vertical integration 1 = Restructuring through vertical separation
	Separation of business lines	0 = No separation of different business services 1 = Separation of different business services
	SOE corporatization	0 = No state-owned utility corporatized 1 = At least one utility corporatized
	Existence of regulatory body	0 = No autonomous regulatory body 1 = Autonomous regulatory body
Policy oversight	Tariff approval oversight	0 = Oversight on tariff approval by line ministry 1 = Oversight on tariff approval by a special entity within the ministry, an interministerial committee, or the regulator
	Investment plan oversight	0 = Oversight on investment plans by line ministry 1 = Oversight on investment plans by a special entity within the ministry, an interministerial committee, or the regulator
	Technical standard oversight	0 = Oversight on technical standards by line ministry 1 = Oversight on technical standards by a special entity within the ministry, an interministerial committee, or the regulator
	Regulation monitoring oversight	0 = Oversight on regulation monitoring by line ministry 1 = Oversight on regulatory monitoring by a special entity within the ministry, an interministerial committee, or the regulator
	Dispute arbitration oversight	0 = Oversight on dispute resolution by line ministry 1 = Oversight on dispute resolution by a special entity within the ministry, an interministerial committee, or the regulator

(continued next page)

Table A4.2a   *(continued)*

Subindex	Indicator	Indicator values
Private sector involvement	Private de jure	0 = Private participation forbidden by law
		1 = Private participation allowed by law
	Private de facto	0 = No private participation
		1 = At least a form of private participation
	Private sector management	0 = No private sector involvement or service and works contracts only
		1 = Management contract, affermage, lease, or concession
	Private sector investment	0 = No private sector involvement, service and works contracts, management contract, affermage, or lease
		1 = Concession
	Absence of distressed private sector participation	0 = Canceled, distressed private sector participation
		1 = Operational, concluded, and not renewed private sector participation
	Absence of renegotiation in private sector participation	0 = Renegotiation
		1 = No renegotiation
	Private ownership	0 = No private ownership
		1 = At least a form of greenfield operation/divestiture
	Full privatization of incumbent operator	0 = No privatization or partial privatization
		1 = Full privatization (sales of 51% or more shares)
	Absence of renationalization	0 = Renationalization
		1 = No renationalization

Source: Vagliasindi and Nellis 2010.
Note: SOE = state-owned enterprise.

Table A4.2b Institutional Indicators: Reform, 2007

Country	Legislation — Existence of reform	Legal reform	Unbundling	Restructuring — Separation of business lines	SOE corporatization	Existence of regulatory body	Policy oversight — Regulation monitoring oversight	Dispute arbitration oversight	Tariff approval oversight	Investment plan oversight	Technical standard oversight	Private sector involvement — Private de jure	Private de facto	Private sector management	Private sector investment	Absence of distressed private sector participation	Absence of renegotiation in private sector participation	Private ownership	Full privatization of incumbent operator	Absence of renationalization
Benin	0	0	0	1	1	0	0	0	0	0	0	0	0	0	0	—	—	0	0	—
Burkina Faso	1	1	0	1	1	0	0	0	0	0	0	1	1	0	0	—	—	1	0	—
Cameroon	1	1	0	1	1	1	1	1	1	1	0	1	1	1	1	1	0	0	0	1
Cape Verde	1	1	0	0	1	:	1	1	1	1	1	1	1	1	1	0	0	0	0	0
Chad	1	1	0	1	1	0	—	—	0	0	—	0	0	0	0	—	—	0	0	—
Congo, Dem. Rep.	0	0	0	1	1	0	0	1	1	1	1	0	0	0	0	—	—	0	0	—
Côte d'Ivoire	1	1	0	1	1	1	0	1	1	1	1	1	1	1	1	1	0	1	0	1
Ethiopia	1	1	1	1	1	0	1	1	1	1	1	0	0	0	0	—	—	0	0	1
Ghana	1	1	1	1	1	1	1	1	1	1	1	1	1	1	1	—	0	1	0	—
Kenya	1	1	1	1	1	1	1	0	1	1	—	0	0	0	0	—	—	0	0	—
Lesotho	—	1	0	0	1	1	1	1	1	—	—	1	1	1	1	1	0	1	0	—
Madagascar	1	1	0	1	1	0	—	1	1	0	1	1	1	1	0	1	—	0	0	—
Malawi	1	1	0	1	1	1	1	1	1	0	1	0	0	0	0	1	0	0	0	—
Mozambique	1	1	0	1	1	0	0	0	0	0	0	1	1	0	0	—	—	0	0	—

Country															
Namibia	1	1	1	1	1	1	1	1	1	0	0	—	—	0	0	—
Niger	1	1	0	1	0	1	0	1	0	0	0	—	—	0	0	—
Nigeria	1	1	1	1	0	1	1	1	1	1	1	—	0	1	1	—
Rwanda	1	1	0	1	0	1	1	1	1	1	0	0	0	0	0	—
Senegal	1	1	0	1	0	1	0	0	0	0	0	—	—	0	1	—
South Africa	1	1	1	1	1	1	1	1	1	1	1	—	—	0	0	—
Sudan	0	—	—	—	—	—	—	—	—	1	1	—	—	—	—	—
Tanzania	1	1	0	1	0	1	1	1	0	1	0	1	1	0	0	—
Uganda	1	1	1	1	1	1	1	1	1	1	1	1	1	0	1	—
Zambia	1	1	0	1	0	1	0	0	0	0	0	—	—	0	0	—

Source: Vagliasindi and Nellis 2010.
Note: SOE = state-owned enterprise; — = data not available.

Table A4.3a   Institutional Indicators: Description of Reform Sector–Specific Indicators

Subindex	Indicator	Indicator values
Restructuring	De jure unbundling of generation and transmission	0 = Joint ownership allowed by law 1 = No joint ownership allowed by law
	De jure unbundling of distribution and transmission	0 = Joint ownership allowed by law 1 = No joint ownership allowed by law
	De jure unbundling of generation and distribution	0 = Joint ownership allowed by law 1 = No joint ownership allowed by law
	De facto unbundling of generation and transmission	0 = No unbundling 1 = Unbundling
	De facto unbundling of distribution and transmission	0 = No unbundling 1 = Unbundling
	De facto unbundling of generation and distribution	0 = No unbundling 1 = Unbundling
Decentralization	Responsibility for urban electricity service provision	0 = No decentralization of responsibility of service provision at the national or state level 1 = Decentralization of responsibility of service provision beyond national or state level
	Responsibility for rural electricity service provision	0 = Accountability of service provision only at the central government level 1 = Accountability of service provision at the regional, state, or local government
Market structure	Single-buyer model	0 = Vertically integrated structure 1 = Single-buyer model
	Separation of water and electricity service provision	0 = No separation of water and electricity service provision 1 = Separation of water and electricity service provision
	De jure IPP	0 = IPPs not allowed by law 1 = IPPs allowed by law
	De facto IPP	0 = No IPPs 1 = At least one IPP
	Community-based providers of rural electricity	0 = No presence of community-based service providers 1 = Presence of community-based service providers

Source: Vagliasindi and Nellis 2010.
Note: IPP = independent power project.

Table A4.3b   Institutional Indicators: Reform Sector Specific, 2007

Country	Restructuring						Market structure		
	De jure unbundling of generation and transmission	De jure unbundling of distribution and transmission	De jure unbundling of generation and distribution	De facto unbundling of generation and transmission	De facto unbundling of distribution and transmission	De facto unbundling of generation and distribution	Single buyer model	De jure IPP	De facto IPP
Benin	0	0	0	1	1	—	0	0	0
Burkina Faso	1	1	1	1	1	1	0	1	1
Cameroon	1	1	1	1	1	1	0	1	0
Cape Verde	0	0	0	1	1	1	0	1	0
Chad	0	0	0	1	1	1	0	1	0
Congo, Dem. Rep.	0	0	0	1	1	1	0	1	0
Côte d'Ivoire	1	1	1	1	1	1	0	1	1
Ethiopia	0	0	0	1	1	1	0	1	0
Ghana	0	0	0	1	1	1	0	1	1
Kenya	1	0	1	0	1	0	1	1	1
Lesotho	—	—	—	—	—	—	—	—	—
Madagascar	0	0	0	1	1	1	0	1	0
Malawi	0	0	0	1	1	1	0	1	0
Mozambique	0	0	0	1	1	1	0	1	0
Namibia	0	0	0	1	0	0	1	1	0
Niger	1	1	1	1	1	1	0	0	0
Nigeria	0	0	0	1	1	1	0	1	1
Rwanda	0	0	0	0	0	0	0	1	0
Senegal	0	0	0	1	1	1	0	1	1
South Africa	0	0	0	1	1	1	0	1	0
Sudan	—	—	—	—	—	—	—	—	—
Tanzania	0	0	0	1	1	1	0	1	1
Uganda	1	1	1	0	0	0	1	1	1
Zambia	0	0	0	1	1	1	0	1	1

Source: Vagliasindi and Nellis 2010.
Note: IPP = independent power project; — = data not available.

Table A4.4a   Institutional Indicators: Description of Regulation Indicators

Subindex	Indicator	Indicator values
Autonomy	Formal autonomy: hire	0 = Appointment by government/line ministry 1 = Otherwise
	Formal autonomy: fire	0 = Firing by government/line ministry 1 = Otherwise
	Partial financial autonomy	0 = Budget fully funded by government 1 = At least a proportion of budget funded through fees and/or donors
	Full financial autonomy	0 = At least a proportion of budget funded through government and/or donors 1 = Budget fully funded through fees
	Partial managerial autonomy	0 = Veto decision by government/line ministry 1 = Veto decision by others
	Full managerial autonomy	0 = Veto decision by government/line ministry/others 1 = No veto decision
	Multisectoral	0 = Sector-specific regulator 1 = Multisectoral regulator
	Commissioner	0 = Individual 1 = Board of Commissioners
Transparency	Publicity of decisions reports only	0 = Regulatory decisions not publicly available 1 = Regulatory decisions publicly available only through reports
	Publicity of decisions Internet only	0 = Regulatory decisions not publicly available or available only through reports 1 = Regulatory decisions publicly available through Internet
	Publicity of decisions public hearing only	0 = Regulatory decisions not publicly available or available through only through reports/Internet 1 = Regulatory decisions publicly available through public hearings
Accountability	Appeal	0 = No right to appeal regulatory decisions 1 = Right to appeal regulatory decision
	Partial independence of appeal	0 = Appeal to government/line ministries 1 = Appeal to bodies other than government/line ministries
	Full independence of appeal	0 = No recourse to independent arbitration 1 = Possibility to appeal to independent arbitration

(continued next page)

Table A4.4a   *(continued)*

Subindex	Indicator	Indicator values
Tools	Tariff methodology	0 = No tariff methodology
		1 = Some tariff methodology (price cap or ROR)
	Tariff indexation	0 = No tariff indexation
		1 = Some tariff indexation
	Regulatory review	0 = No tariff review
		1 = Periodic tariff review
	Length of regulatory review	0 = No tariff review, annual review or review (period less than three years)
		1 = Multiyear tariff review (at least three years)
USO	Existence of a specific sectoral fund	0 = No sectoral fund established
		1 = Sectoral fund established
	Finance of sectoral fund	0 = No funding based on levies
		1 = At least a percentage of funding coming through sectoral levies

Source: Vagliasindi and Nellis 2010.
Note: ROR = rate of return; USO = universal service obligation.

Table A4.4b   Institutional Indicators: Regulation, 2007

	Autonomy							
	Formal autonomy hire	Formal autonomy fire	Partial financial autonomy	Full financial autonomy	Partial managerial autonomy	Full managerial autonomy	Multisectoral	Commissioner
Benin	—	—	—	—	—	—	—	—
Burkina Faso	—	—	—	—	—	—	—	—
Cameroon	0	0	1	0	0	0	0	1
Cape Verde	0	1	1	1	0	0	1	1
Chad	—	—	—	—	—	—	—	—
Congo, Dem. Rep.	—	—	—	—	—	—	—	—
Côte d'Ivoire	0	0	1	1	—	1	0	0
Ethiopia	0	0	1	0	0	0	0	1
Ghana	0	0	—	—	—	—	1	1
Kenya	0	0	1	1	0	0	0	1
Lesotho	0	0	1	—	0	0	0	1
Madagascar	0	0	1	1	0	0	0	1
Malawi	0	0	1	0	1	0	0	1
Mozambique	—	—	—	—	—	—	—	—
Namibia	0	0	1	1	0	0	0	1
Niger	0	0	1	0	0	0	1	0
Nigeria	0	0	1	0	1	0	0	1
Rwanda	1	0	1	1	0	0	1	1
Senegal	0	0	—	—	1	0	0	1
South Africa	0	0	1	1	1	0	1	1
Sudan	—	—	—	—	—	—	—	—
Tanzania	0	0	1	1	0	0	1	1
Uganda	0	0	1	0	1	0	0	1
Zambia	0	0	1	0	0	0	1	1

Source: Vagliasindi and Nellis 2010.
Note: — = data not available.

Transparency			Accountability			Tools			
Publicity of decisions reports only	Publicity of decisions Internet only	Publicity of decisions public hearing only	Appeal	Partial independence of appeal	Full independence of appeal	Tariff methodology	Tariff indexation	Regulatory review	Length of regulatory review
—	—	—	—	—	—	0	0	0	—
—	—	—	0	—	—	1	—	—	—
0	0	0	1	1	0	1	1	1	1
1	1	1	1	1	0	1	1	1	1
—	—	—	—	—	—	0	—	—	—
—	—	—	—	—	—	0	0	0	—
0	0	—	1	1	0	0	0	1	—
1	1	0	1	1	0	1	0	1	—
1	1	1	—	—	—	1	—	—	—
1	1	0	1	0	0	1	0	1	1
—	—	—	—	—	—	1	—	—	—
1	1	0	1	0	0	1	1	1	1
1	0	1	1	1	0	1	0	1	1
—	—	—	—	—	—	0	—	—	—
1	1	1	1	1	0	1	1	—	—
1	1	1	1	1	0	0	0	—	—
1	1	1	0	—	—	1	0	1	1
1	1	1	0	—	—	1	0	—	—
1	0	0	1	1	0	1	1	1	1
1	1	1	1	1	0	1	1	1	1
—	—	—	—	—	—	—	—	—	—
1	1	1	1	1	0	1	0	1	0
1	1	1	1	1	0	1	1	0	0
1	1	1	1	1	0	1	0	1	0

Table A4.5a Institutional Indicators: Description of Regulation Sector-Specific Indicators

Subindex	Indicator	Indicator values
	Regulation of large customers	0 = Regulation of large customers
		1 = No regulation of large customers
	Transmission tariff regulation methodology	0 = No transmission tariff regulation methodology
		1 = Tariff regulation methodology (price cap or rate of return)
Tools	Third-party access to T&D networks	0 = TPA not allowed by law
		1 = TPA allowed by law
	Minimum quality standards	0 = No well-defined minimum quality standards
		1 = Well-defined minimum quality standards
	Penalties for noncompliance	0 = No penalties for noncompliance to minimum quality standards
		1 = Penalties for noncompliance to minimum quality standards
	Partial cost recovery requirement for rural electricity	0 = Full capital subsidy
		1 = Partial capital subsidy
Cost recovery	Full cost recovery requirement for rural electricity	0 = Partial or full capital subsidy
		1 = No subsidy
	Community contribution to the rural fund	0 = No community contribution to the rural fund
		1 = Community contribution to the rural fund
	Criteria are used to prioritize rural electrification projects	0 = Criteria other than least cost
		1 = Least-cost criteria
Universal service	Opex cost recovery for rural water	0 = No opex recovery
		1 = Opex recovery
	Capex cost recovery for water	0 = Only opex recovery or no recovery
		1 = Some capex recovery
Environmental	Incentives for renewable energy	0 = No incentive for renewable energy
		1 = Incentives for renewable energy

Source: Vagliasindi and Nellis 2010.

Note: capex = capital expenses; opex = operational expenses; T&D = transmission and distribution; TPA = third-party access.

Table A4.5b   Institutional Indicators: Regulation Sector Specific, 2007

Country	Minimum quality standards	Penalties for noncompliance	Cutoff possibility	Criteria used to prioritize rural electrification projects	Regulation of large customers	Transmission tariff regulation methodology	Third-party access to transmission and distribution networks	Partial cost recovery requirement for rural electricity	Full cost recovery requirement for rural electricity	Incentives for renewable energy
	Tools			Access/interconnection				Cost recovery		Environmental
Benin	0	—	1	0	1	—	0	—	0	—
Burkina Faso	0	0	1	0	0	1	0	—	0	—
Cameroon	1	1	1	0	1	1	0	—	—	—
Cape Verde	1	0	1	0	0	1	0	—	0	—
Chad	0	0	1	0	1	—	—	—	0	—
Congo, Dem. Rep.	1	1	1	0	1	—	1	—	—	—
Côte d'Ivoire	1	1	1	0	1	—	0	—	1	—
Ethiopia	1	0	1	0	0	1	0	—	0	—
Ghana	1	—	1	0	0	1	1	—	0	—
Kenya	1	0	1	1	0	1	—	—	1	—
Lesotho	—	—	—	—	—	—	—	—	0	—
Madagascar	1	0	1	0	0	1	—	—	1	—
Malawi	1	1	1	0	0	1	—	—	—	—
Mozambique	—	—	1	0	1	0	—	—	0	—
Namibia	1	1	1	0	0	1	—	—	—	—
Niger	0	0	1	0	1	—	—	—	1	—
Nigeria	1	1	1	0	0	1	—	—	0	—
Rwanda	0	0	1	0	0	1	—	—	1	—
Senegal	1	1	1	0	0	1	—	—	1	—
South Africa	1	1	1	0	0	1	—	—	0	—
Sudan	—	—	—	—	—	—	—	—	—	—
Tanzania	0	0	1	0	0	1	—	—	—	—
Uganda	1	1	0	0	0	1	—	—	0	—
Zambia	1	1	1	—	0	1	—	1	0	—

Source: Vagliasindi and Nellis 2010.
Note: — = data not available.

Table A4.6a   Institutional Indicators: Description of SOE Governance Indicators

Subindex	Indicator	Indicator values
Ownership and shareholder quality	Concentration of ownership	0 = Ownership diversified 1 = 100% owned by one state body
	Corporatization	0 = Noncorporatized 1 = Corporatized
	Limited liability	0 = Nonlimited liability 1 = Limited liability company
	Rate of return policy	0 = No requirement to earn a rate of return 1 = Requirement to earn a rate of return
	Dividend policy	0 = No requirement to pay dividends 1 = Requirement to pay dividends
Managerial and board autonomy	Hiring	0 = Manager does not have the most decisive influence on hiring decisions 1 = Manager has the most decisive influence on hiring decisions
	Laying off	0 = Manager does not have the most decisive influence on firing decisions 1 = Manager has the most decisive influence on firing decisions
	Wages	0 = Manager does not have the most decisive influence on setting wages/bonuses 1 = Manager has the most decisive influence on setting wages/bonuses
	Production	0 = Manager does not have the most decisive influence on how much to produce 1 = Manager has the most decisive influence on how much to produce
	Sales	0 = Manager does not have the most decisive influence on to whom to sell 1 = Manager has the most decisive influence on to whom to sell
	Size of the board	0 = Number of members of board smaller than a given threshold 1 = Number of members of board greater than a given threshold
	Selection of board members	0 = Board members appointed only by government 1 = Board members appointed by shareholders
	Presence of independent directors	0 = No independent directors in the board 1 = At least one independent director in the board

(continued next page)

Table A4.6a   *(continued)*

Subindex	Indicator	Indicator values
	Publication of annual reports	0 = Annual reports not publicly available 1 = Annual reports publicly available
	IFRS	0 = IFRS not applied 1 = Compliance to IFRS
	External audits	0 = No operational or financial audit 1 = At least some form of external audit
Accounting and disclosure and performance monitoring	Independent audit of accounts	0 = No independent audit of accounts 1 = Independent audit of accounts
	Audit publication	0 = Audit not publicly available 1 = Audit publicly available
	Remuneration of noncommercial activity	0 = No remuneration of noncommercial activities 1 = Remuneration of noncommercial activities
	Performance contracts	0 = No performance contracts 1 = Existence of performance contract
	Performance contracts with incentives	0 = Performance-based incentive systems 1 = Existence of performance-based incentive systems
	Penalties for poor performance	0 = No penalties for poor performance 1 = Penalties for poor performance
	Monitoring	0 = No periodic monitoring of performance 1 = Periodic monitoring of performance
	Third-party monitoring	0 = No monitoring of performance by third party 1 = Monitoring of performance by third party
Outsourcing	Billing and collection	0 = Ownership diversified 1 = 100% owned by one state body
	Meter reading	0 = Noncorporatized 1 = Corporatized
	Human resources	0 = Nonlimited liability 1 = Limited liability company
	Information technology	0 = No requirement to earn a rate of return 1 = Requirement to earn a rate of return

(continued next page)

Table A4.6a   *(continued)*

Subindex	Indicator	Indicator values
Labor market discipline	Restriction to dismiss employees	0 = Restrictions to dismiss employees according to public service guidelines 1 = Restrictions to dismiss employees only according to corporate law
	Wages compared with private sector	0 = Wages compared with public sector 1 = Wages compared with private sector (or between public and private sectors)
	Benefits compared with private sector	0 = Benefits compared with public sector 1 = Benefits compared with private sector (or between public and private sectors)
Capital market discipline	No exemption from taxation	0 = Exemption from taxation 1 = No exemption from taxation
	Access to debt compared with private sector	0 = Access to debt below the market rate 1 = Access to debt equal or above the market rate
	No state guarantees	0 = At least one state guarantee 1 = No state guarantee
	Public listing	0 = No public listing 1 = Public listing

Source: Vagliasindi and Nellis 2010.
Note: IFRS = International Financial Reporting Standards.

Table A4.6b Institutional Indicators: SOE Governance, 2007

Country	Provider name	Concentration of ownership	Corporatization	Limited liability	Rate of return policy	Dividend policy	Hiring	Laying off	Wages	Production	Sales	Size of the board	Selection of board members	Presence of independent directors
Benin	SBEE	1	—	—	0	0	1	1	1	0	1	1	0	—
Burkina Faso	Sonabel	1	1	0	1	1	1	1	1	1	1	1	—	—
Cameroon	AES SONEL	0	1	1	1	1	1	1	1	0	0	1	0	—
Cape Verde	Electra	1	1	0	1	0	1	1	1	0	1	1	1	1
Chad	STEE	1	0	0	1	1	1	1	1	1	1	1	1	1
Congo, Dem. Rep.	SNEL	0	—	—	0	1	1	1	1	—	—	1	0	—
Côte d'Ivoire	CIE	1	—	0	1	0	1	1	1	1	1	1	1	—
Ethiopia	EEPCO	1	—	—	—	—	—	—	1	—	—	—	—	—
Ghana	ECG	1	1	1	1	1	1	0	1	1	1	1	1	—
Ghana	VRA	0	1	1	1	1	1	1	1	1	—	1	1	—
Kenya	KENGEN	0	1	1	1	1	1	1	—	1	0	1	1	—
Kenya	KPLC	1	1	0	—	0	1	1	1	0	1	1	0	—
Lesotho	LEC	1	0	0	—	—	1	1	1	1	0	1	1	—
Madagascar	JIRAMA	1	—	—	1	1	1	1	1	1	1	1	1	—
Malawi	Escom	1	—	—	1	1	1	1	1	1	1	—	—	—
Mozambique	EDM	1	1	—	—	—	—	—	—	—	—	—	—	—

(continued next page)

Table A4.6b (continued)

Country	Provider name	Ownership and shareholder quality					Managerial and board autonomy							
		Concentration of ownership	Corporatization	Limited liability	Rate of return policy	Dividend policy	Hiring	Laying off	Wages	Production	Sales	Size of the board	Selection of board members	Presence of independent directors
Namibia	Nampower	—	—	—	—	—	1	—	1	1	0	1	1	—
	NORED	1	—	—	0	1	1	1	1	1	1	1	0	1
Niger	NiGELEC	1	1	—	1	—	1	1	1	1	1	1	0	—
Nigeria	PHNC	1	0	0	1	0	1	1	1	1	1	0	1	—
Rwanda	ELECTROGAZ	1	0	0	0	0	1	1	1	1	0	1	0	—
Senegal	Senelec	—	—	—	1	1	1	1	0	1	0	—	1	—
South Africa	Capetown	1	1	1	1	1	1	1	1	1	0	1	1	—
	ESKOM	—	1	—	—	—	—	—	—	1	—	—	—	—
	Tshwane	1	1	—	—	—	—	—	—	1	—	—	—	—
Sudan	NEC	1	1	—	—	—	—	—	—	—	—	—	1	1
Tanzania	TANESCO	0	1	0	0	0	1	1	1	1	1	0	0	—
Uganda	UEDCL	0	1	0	0	0	1	1	1	0	0	0	1	—
	UEGCL	1	1	0	0	0	1	1	1	1	0	0	0	—
	UETCL	1	—	—	1	1	1	1	1	1	1	1	1	—
Zambia	ZESCO	0	—	—	0	0	1	1	1	0	—	1	1	—

(continued next page)

Table A4.6b (continued)

Country	Accounting and disclosure and performance monitoring											Outsourcing				Labor market discipline		Capital market discipline				
	Publication of annual reports	IFRS	External audits	Independent audit of accounts	Audit publication	Remuneration of noncommercial activity	Performance contracts	Performance contracts with incentives	Penalties for poor performance	Monitoring	Third-party monitoring	Billing and collection	Meter reading	Human resources (HR)	IT	Restriction to dismiss employees	Wages compared to private sector	Benefits vs. private sector	No exemption from taxation	Access to debt vs. private sector	No state guarantees	Public listing
Benin	○	1	1	1	○	○	○	1	○	○	1	○	—	○	○	—	1	—	1	○	○	—
Burkina Faso	1	1	1	1	○	○	1	○	—	—	—	1	—	1	1	—	1	—	○	○	○	○
Cameroon	1	1	1	—	○	○	1	1	○	○	—	1	—	1	1	—	1	—	—	—	—	○
Cape Verde	—	○	1	—	1	○	—	○	—	—	—	○	—	○	○	—	1	—	—	1	○	○
Chad	1	1	1	○	○	○	○	○	1	—	—	○	—	○	○	—	1	—	—	—	1	○
Congo, Dem. Rep.	○	1	1	—	—	○	○	○	○	○	—	○	—	○	○	—	1	—	—	—	○	○
Côte d'Ivoire	1	1	1	1	1	○	○	1	1	○	—	○	—	○	○	—	—	—	—	—	○	○
Ethiopia	—	—	—	—	—	—	—	—	—	—	—	—	—	—	—	—	—	—	—	—	—	—
Ghana	1	○	1	1	1	○	1	1	○	—	—	○	—	○	○	—	—	—	○	—	○	○
Kenya	1	1	1	—	—	○	—	—	—	1	1	—	—	—	—	—	1	—	—	1	○	○
Lesotho	○	○	—	—	—	○	1	1	○	○	—	○	—	○	○	—	1	—	○	—	○	○
Madagascar	○	1	1	1	○	○	○	○	○	○	—	○	—	○	—	—	1	—	—	1	1	○
Malawi	1	1	1	—	—	○	1	○	1	1	—	○	—	1	1	—	1	—	1	—	—	○
Mozambique	—	—	1	1	1	○	—	—	—	○	—	○	—	—	○	—	1	—	—	—	○	—
Namibia	○	1	1	1	○	○	○	○	—	○	—	○	—	1	1	—	1	—	1	—	○	—

Table A4.6b (continued)

Country	Accounting and disclosure and performance monitoring — Publication of annual reports	IFRS	External audits	Independent audit of accounts	Audit publication	Remuneration of noncommercial activity	Performance contracts	Performance contracts with incentives	Penalties for poor performance	Monitoring	Third-party monitoring	Outsourcing — Billing and collection	Meter reading	Labor market discipline — Human resources (HR)	IT	Restriction to dismiss employees	Wages compared to private sector	Capital market discipline — Benefits vs. private sector	No exemption from taxation	Access to debt vs. private sector	No state guarantees	Public listing
Niger	1	1	1	1	1	0	1	1	1	1	1	1	—	0	0	—	1	—	0	1	0	—
Nigeria	0	0	1	1	0	0	0	0	0	0	1	0	—	0	1	—	1	—	1	1	0	0
Rwanda	0	1	1	1	0	0	0	0	0	—	1	0	—	0	0	—	1	—	1	0	0	0
Senegal	1	0	1	1	1	0	1	1	1	1	1	0	—	0	0	—	1	—	0	0	1	—
South Africa	1	1	—	—	1	0	1	1	1	1	—	0	—	0	1	—	1	—	1	1	1	0
Sudan	—	—	—	—	—	0	—	—	—	—	—	—	—	—	—	—	—	—	—	—	—	—
Tanzania	1	1	1	1	1	0	0	1	0	1	1	0	—	0	1	—	1	—	1	1	1	0
Uganda	1	1	1	1	1	0	1	1	1	—	1	0	—	0	0	—	1	—	1	1	0	0
Zambia	1	0	1	1	1	0	0	0	0	1	1	—	—	0	0	—	1	—	1	—	0	—

Source: Vagliasindi and Nellis 2010.

Note: IFRS = International Financial Reporting Standards; — = data not available.

Table A4.7　Private Participation: Greenfield Projects, 1990–2006

Country	Project name	Technology/fuel	Project type	Capacity year	Contract period	Termination year	Investment in facilities ($ million)	MW developed
Angola	Chicapa Hydroelectric Plant	Hydro	Build, own, transfer	2003	40	2044	45	16
	Aggreko Cabinda Temporary Power Station	Diesel	Rental	2006	2	2009	4.7	30
	Aggreko Caminhos de Ferro de Angola	Diesel	Rental	2006	2	2009	4.7	30
Burkina Faso	Hydro-Afrique Hydroelectric Plant	Hydro	Build, own, transfer	1998	14	2012	5.6	12
Congo, Rep.	Sounda S.A.	Geothermal	Build, own, operate	1996	—	1998	325	240
	Compagnie Ivoirienne de Production d'Electricité (CIPREL)	Diesel, natural gas	Build, own, transfer	1994	19	2013	70	99
Ghana	Azito Power Project	Natural gas	Build, own, transfer	1999	23	2022	223	420
	SIIF Accra	Steam	Merchant	1999	2	2001	0	39
	Takoradi 2	Natural gas	Build, own, operate	1999	25	2024	110	220
Kenya	Iberafrica Power Ltd.	Diesel	Build, own, transfer	1996	7	2011	64	56
	Mombasa Barge-Mounted Power Project	Diesel	Build, own, operate	1996	7	2004	35	46
	Kipevu II	Diesel	Build, own, operate	1999	20	2019	85	75
	Ormat Olkaria III Geothermal Power Plant (phase 1)	Geothermal	Build, own, operate	1999	20	2019	54	13
	Aggreko Embakassi and Eldoret Power Stations	—	Rental	2006	1	2007	7.89	100
Mauritius	Deep River Beau Champ	Coal, waste	Build, own, operate	1998	—	—	0	29
	Belle Vue Power Plant	Coal	Build, own, operate	1998	20	2018	109.3	100
	FUEL Power Plant	Coal, waste	Build, own, operate	1998	20	2018	0	40
	St. Aubin Power Project	Coal, waste	Build, own, operate	2004	20	2025	0	34

(continued next page)

Table A4.7 *(continued)*

Country	Project name	Technology/fuel	Project type	Capacity year	Contract period	Termination year	Investment in facilities ($ million)	MW developed
Nigeria	AES Nigeria Barge Limited	Natural gas	Build, own, operate	2001	—	—	240	306
	Okpai Independent Power Project	Natural gas	Build, own, operate	2002	—	—	462	450
	AEL Ilorin Gas Power Plant	Natural gas	Build, own, operate	2005	—	—	275	105
	Dadin Kowa Hydropower Plant	Hydro	Build, own, transfer	2005	25	2030	26	39
Rwanda	Aggreko 10 MW Power Station Rwanda	—	Rental	2005	2	2007	1.58	10
Senegal	GTi Dakar Ltd.	Diesel, natural gas	Build, own, transfer	1997	15	2012	59	56
	Aggreko Dakar Temporary Power Station	—	Rental	2005	2	2008	6.31	40
	Kounoune I IPP	Diesel	Build, own, operate	2005	15	2020	87	68
South Africa	Bethlehem Hydro	Hydro	Build, own, operate	2005	20	2025	7	4
	Darling Wind Farm	Wind	Build, own, operate	2006	20	2026	9.9	5
Tanzania	Tanwat Wood-Fired Power Plant	Waste	Build, lease, own	1994	6	2000	6	2.5
	Independent Power Tanzania Ltd	Diesel	Build, own, transfer	1997	20	2017	127	100
	Songas–Songo Songo Gas-to-Power Project	Natural gas	Build, own, transfer	2001	20	2021	316	—
	Songas–Songo Songo Gas-to-Power Project	Natural gas	Build, own, transfer	2004	20	2021	0	115
	Songas–Songo Songo Gas-to-Power Project	Natural gas	Build, own, transfer	2005	20	2021	0	190
	Mtwara Region Gas-to-Power Project	Natural gas	Build, own, operate	2005	25	2021	32	12
	Aggreko Ubungo Temporary Power Station	Natural gas	Rental	2006	2	2009	6.31	40
	Alstom Power Rentals Mwanza	Diesel	Rental	2006	2	2008	6.31	40
	Dowans Lease Power Ubungo	Natural gas	Rental	2006	2	2009	15.78	100
Uganda	Aggreko Kampala Temporary Power Station	Diesel	Rental	2005	3	2008	11.83	50
	Aggreko Jinja Temporary Power Station	Diesel	Rental -	2006	2	2008	11.8	50

Source: World Bank 2007.

Note: Termination year can be year when the project is concluded according to the original agreement, rescheduling, or project cancellation. MW = megawatts; — = data not available.

Table A4.8 Private Participation: Concessions, Management and Lease Contracts, Divestitures, 1990–2006

Country	Project name	Project type	Project status	Capacity year	Contract period	Termination year	% private	Investment in facilities ($ million)	Number of connections (thousands)	MW
Concession:										
Cameroon	AES Sonel	Build, rehab., operate, transfer	Operational	2001	20	2021	56	39.8	452	—
	AES Sonel	Build, rehab., operate, transfer	Operational	2002	20	2021	56	21.5	—	—
	AES Sonel	Build, rehab., operate, transfer	Operational	2005	20	2021	56	0	528	n.a.
	AES Sonel	Build, rehab., operate, transfer	Operational	2006	20	2021	56	440	528	n.a.
Comoros	Comorienne de d'eau et de l'electricité (CEE)	Rehab, operate, transfer	Canceled	1998	n.a.	2001	100	0	n.a.	16
Côte d'Ivoire	Compagnie Ivoirienne d'Electricité	Rehab, operate, transfer	Operational	1990	20	2010	100	39.6	411.7	n.a.
	Compagnie Ivoirienne d'Electricité	Rehab, operate, transfer	Operational	2000	20	2010	100	0	760	n.a.
	Compagnie Ivoirienne d'Electricité	Rehab, operate, transfer	Operational	2005	20	2010	85	0	—	—

(continued next page)

Table A4.8 *(continued)*

Country	Project name	Project type	Project status	Capacity year	Contract period	Termination year	% private	Investment in facilities ($ million)	Number of connections (thousands)	MW
Gabon	Société d'Energie et d'Eau du Gabon (SEEG)	Build, rehab, operate, transfer	Operational	1997	20	2017	100	268	84	n.a.
	Société d'Energie et d'Eau du Gabon (SEEG)	Build, rehab, operate, transfer	Operational	2002	20	2017	100	0	125	n.a.
Guinea	Société Guineenne d'Electricité	Rehab, lease or rent, transfer	Concluded	1995	10	2005	66	36.4	n.a.	180
Mali	Energie du Mali (EDM)	Build, rehab, operate, transfer	Distressed	2000	26	2020	60	337	120	n.a.
	Energie du Mali (EDM)	Build, rehab, operate, transfer	Distressed	2004	26	2020	34	0	—	—
Mozambique	Energia de Mocambique Lda (ENMo)	Build, rehab, operate, transfer	Operational	2004	20	2024	100	5.8	3000	n.a.
Nigeria	Afam Power Project	Rehab, operate, transfer	Operational	2005	15	2020	100	238	n.a.	400
São Tomé and Príncipe	Sinergie concession contract	Build, rehab, operate, transfer	Operational	2004	45	2049	100	50	n.a.	—

Senegal	Société Nationale d'Electricité du Senegal (SENELEC)	Build, rehab., operate, transfer	Canceled	1999	25	2000	34	0	n.a.	300
South Africa	PN Energy Services (Pty) Ltd	Build, rehab., operate, transfer	Operational	1995	n.a.	n.a.	67	3	—	n.a.
Togo	Togo Electricité	Rehab., lease or rent, transfer	Canceled	2000	20	2006	100	36	—	—
Uganda	Kasese Electrification Project	Rehab., operate, transfer	Operational	2003	n.a.	n.a.	n.a.	0	n.a.	5.5
	Uganda Electricity Generation Company Limited	Rehab., lease or rent, transfer	Operational	2003	20	2023	100	6.8	n.a.	300
	Western Nile Rural Electrification Project	Rehab., operate, transfer	Operational	2003	20	2023	100	11.3	n.a.	3.5
	Umeme Limited	Rehab., lease or rent, transfer	Operational	2005	20	2025	100	65	250	n.a.
Management and lease contracts:										
Chad	Société Tchadienne d'Eau et d'Electricité (STEE)		Canceled	2000	30	2004	100	0	16	n.a.
Gabon	Société Africaine de Gestion et d'Investissement (SAGI)	Management contract	Concluded	1993	4	1997	100	0	—	—

(continued next page)

Table A4.8 *(continued)*

Country	Project name	Project type	Project status	Capacity year	Contract period	Termination year	% private	Investment in facilities ($ million)	Number of connections (thousands)	MW
Gambia, The	Management Service Gambia (MSG)	Lease contract	Canceled	1993	10	1996	100	0	—	—
	National Water and Electricity Company Management Contract	Management contract	Operational	2006	5	2011	100	0	n.a.	40
Ghana	Electricity Corporation of Ghana	Management contract	Concluded	1994	4	1998	100	0	500	n.a.
Guinea-Bissau	Electricidade e Aguas de Guinea-Bissau	Management contract	Concluded	1991	4	1997	100	0	n.a.	10.4
Kenya	Kenya Power and Lighting Company Management Contract	Management contract	Operational	2006	2	2008	100	0	800	n.a.
Lesotho	Lesotho Electricity Corporation (LEC)	Management contract	Operational	2002	n.a.	n.a.	100	0	n.a.	—

Madagascar	Jiro sy Rano Malagasy (Jirama)	Management contract	Operational	2005	2	2007	n.a.	0	340	n.a.
Malawi	Electricity Supply Corporation of Malawi Ltd (ESCOM)	Management contract	Concluded	2001	2	2003	100	0	n.a.	300
Mali	Electricité et Eau du Mali (Management)	Management contract	Concluded	1994	5	2000	100	0	—	—
Namibia	Northern Electricity	Management and lease contract	Concluded	1996	5	2002	n.a.	4	n.a.	—
Namibia	Reho-Electricity	Lease contract	Operational	2000	n.a.	n.a.	n.a.	1	—	n.a.
Rwanda	ELECTROGAZ	Management contract	Canceled	2003	5	2006	100	0	25	n.a.
	ELECTROGAZ	Management contract	Canceled	2005	5	2006	100	0	67	n.a.
São Tomé and Príncipe	Empresa de Agua e Electricidade	Management contract	Concluded	1993	3	1996	100	0	n.a.	4.75
Tanzania	Tanzania Electricity Supply Company (TANESCO)	Management contract	Concluded	2002	4	2006	0	0	—	—
Togo	Companie Energie Electrique du Togo	Management contract	Concluded	1997	4	2000	n.a.	0	n.a.	n.a.

(continued next page)

Table A4.8 *(continued)*

Country	Project name	Project type	Project status	Capacity year	Contract period	Termination year	% private	Investment in facilities ($ million)	Number of connections (thousands)	MW
Divestitures:										
Cape Verde	Electra	Partial	Operational	1999	30	2030	51	0	35	n.a.
Cape Verde	Electra	Partial	Operational	2003	30	2030	51	0	71	n.a.
Kenya	Kenya Electricity Generating Company	Partial	Operational	2006	n.a.	n.a.	30	0	n.a.	945
South Africa	AES Kelvin Power	Partial	Operational	2001	20	2021	50	28.4	n.a.	600
Zambia	Consolidated Copper Mines Ltd. Power Division distribution	Partial	Operational	1997	n.a.	n.a.	80	92.5	n.a.	n.a.
	Lunsemfwa Hydro Power	Full	Operational	2001	n.a.	n.a.	100	3	n.a.	38
Zimbabwe	African Power	Partial	Operational	1998	n.a.	n.a.	51	0	n.a.	920

Source: World Bank 2007.

Note: Termination year can be year when the project is concluded according to the original agreement, rescheduling, or project cancellation. MW = megawatts; n.a. = not applicable; — = data not available.

APPENDIX 5

Widening Connectivity and Reducing Inequality

Table A5.1 Access to Electricity
percentage of population

Country	By time period (national)			By location		By expenditure quintile				
	Early 1990s	Late 1990s	Early 2000s	Rural	Urban	Q1	Q2	Q3	Q4	Q5
Benin	—	14	22	6	51	0	1	3	24	82
Burkina Faso	6	6	10	1	54	0	0	1	2	57
Cameroon	31	42	46	16	77	1	14	37	78	98
Central African Republic	5	—	—	1	11	0	0	0	1	25
Chad	—	3	4	0	20	0	0	0	0	21
Comoros	—	30	—	21	54	0	7	17	48	84
Congo, Dem. Rep.	—	—	—	—	—	—	—	—	—	—
Congo, Rep.	—	—	35	16	51	5	14	20	47	88
Côte d'Ivoire	39	50	—	27	90	4	19	41	87	100
Ethiopia	—	11	12	2	86	0	0	1	3	56
Gabon	—	75	—	31	91	17	69	93	98	99
Ghana	28	39	44	21	77	8	39	28	57	90
Guinea	—	17	21	3	63	0	0	4	18	83
Kenya	9	12	13	4	51	0	0	1	7	57
Lesotho	—	—	6	1	28	0	0	0	1	27
Madagascar	9	11	19	10	52	0	0	1	11	82
Malawi	4	6	7	2	34	0	1	0	3	34
Mali	—	8	13	3	41	1	3	2	5	54
Mauritania	—	—	23	3	51	0	2	5	29	81
Mozambique	—	10	11	1	30	0	0	1	4	51
Namibia	20	32	—	10	75	1	1	6	51	100

Niger	6	8	—	0	41	0	0	0	4	36
Nigeria	26	45	51	35	84	10	37	40	78	91
Rwanda	2	7	5	1	27	0	0	1	1	25
Senegal	25	32	46	19	82	4	12	46	76	94
South Africa	—	63	—	36	86	10	36	74	98	100
Sudan	—	—	—	—	—	—	—	—	—	—
Tanzania	6	7	11	2	39	0	0	0	3	50
Togo	—	15	—	2	44	0	0	2	10	62
Uganda	7	—	8	3	47	0	0	2	2	38
Zambia	23	20	20	3	50	0	0	0	15	84
Zimbabwe	23	34	—	7	90	0	12	12	50	97
Overall	23	28	31	12	71	4	14	20	38	72
Income group										
Low income	17	24	27	11	69	3	12	15	32	68
Middle income	59	55	53	27	81	7	28	59	86	97
Urbanization										
Low	8	8	11	3	56	0	0	1	4	52
Medium	24	30	28	3	48	0	1	2	11	60
High	37	47	51	30	83	8	32	45	79	94
Region										
East	17	24	27	2	60	0	0	1	3	53
West	25	37	43	20	78	6	23	27	55	80
South	36	37	35	13	66	4	14	28	42	77
Central	25	28	29	8	66	1	11	20	47	76

Source: Banerjee and others 2008.
Note: Location and expenditure quintile data are for the latest available year. — = data not available.

Table A5.2 Adjusted Access, Hook-Up, Coverage of Electricity, Latest Available Year, Urban Areas

Country	Access	Hookup	Coverage	Unserved population	Pure demand-side gap	Supply-side gap	Pure supply-side gap	Mixed demand- and supply-side gap	Share of deficit attributable to demand-side factors only	Share of deficit attributable to supply-side factors only	Share of deficit attributable to both supply- and demand-side factors
Benin	83	83	51	49	18	31	26	5	36	53	11
Burkina Faso	92	62	54	46	3	43	27	16	7	57	35
Cameroon	94	98	77	23	15	8	8	0	65	34	1
Central African Republic	57	33	11	89	8	81	27	54	9	30	61
Chad	77	28	20	80	1	79	22	57	2	27	71
Comoros	100	82	54	46	28	18	15	3	61	32	7
Congo, Rep.	98	86	51	49	33	16	13	2	68	27	4
Côte d'Ivoire	100	99	90	10	10	1	1	0	93	7	0
Ethiopia	99	90	86	14	3	11	10	1	20	72	8
Gabon	100	99	91	9	9	1	1	0	94	6	0
Ghana	98	95	77	23	15	8	7	0	67	31	2
Guinea	89	78	63	37	6	31	24	7	16	66	18
Kenya	80	72	51	49	6	42	31	12	13	63	24
Lesotho	87	42	28	72	8	64	27	37	12	37	51
Madagascar	80	86	52	48	16	32	27	5	34	56	10
Malawi	84	48	34	66	7	59	29	31	10	43	47
Mali	81	62	41	59	9	49	31	19	16	52	32

Mauritania	85	84	51	49	20	29	24	5	41	49	10
Mozambique	80	51	30	70	11	59	30	29	16	43	41
Namibia	93	99	75	25	17	8	8	0	69	31	0
Niger	94	49	41	59	6	54	26	27	9	45	46
Nigeria	98	98	84	16	12	4	4	0	73	26	1
Rwanda	72	46	27	73	6	67	31	36	8	42	49
Senegal	99	100	82	18	17	1	1	0	95	5	0
South Africa	95	100	86	14	8	6	6	0	58	42	0
Tanzania	83	55	39	61	7	54	30	25	11	49	40
Togo	96	66	44	56	19	37	24	12	34	44	22
Uganda	93	58	47	53	6	46	27	20	12	51	37
Zambia	84	84	50	50	21	29	24	5	42	49	9
Zimbabwe	100	99	90	10	8	2	1	0	85	15	0
Overall	93	87	71	29	11	18	12	6	52	37	11
Income group											
Low income	93	84	69	31	11	21	13	7	50	37	13
Middle income	95	98	81	19	11	7	7	1	61	38	1
Urbanization											
Low	87	67	53	47	6	41	24	17	15	56	29
Medium	86	73	52	48	13	35	22	14	37	42	21
High	97	98	83	17	12	5	5	0	71	28	1
Region											
East	89	71	59	41	5	36	23	14	15	60	26
West	96	93	78	22	12	10	8	2	67	29	5
South	90	87	69	31	10	20	14	7	48	41	10
Central	69	80	60	40	15	25	12	13	53	30	17

Source: Banerjee and others 2008.

Table A5.3 Electricity Expenditure and Its Share in Household Budget

Country	Year	Expenditure budget (2002 $)								Share in household budget (%)							
		National[a]	Rural	Urban	Q1	Q2	Q3	Q4	Q5	National	Rural	Urban	Q1	Q2	Q3	Q4	Q5
Angola	2000	—	—	—	—	—	—	—	—	—	—	—	—	—	—	—	—
Benin	2002	—	—	—	—	—	—	—	—	—	—	—	—	—	—	—	—
Burkina Faso	2003	4	11	1	12	10	8	6	2	3	10	1	25	15	9	5	1
Burundi	1998	—	—	—	—	—	—	—	—	—	—	—	—	—	—	—	—
Cameroon	2004	1	1	1	1	0	0	1	2	1	1	1	3	0	0	1	1
Cape Verde	2001	2	2	3	2	2	2	2	3	2	2	1	3	2	2	2	1
Chad	2001	13	5	15	3	5	8	12	18	4	3	3	3	4	4	4	3
Congo, Dem. Rep.	2005	2	1	2	1	1	1	2	3	2	1	2	2	2	2	2	2
Congo, Rep.	2002	—	—	—	—	—	—	—	—	—	—	—	—	—	—	—	—
Côte d'Ivoire	2005	7	5	7	3	5	7	7	9	3	3	3	3	3	4	3	2
Ethiopia	2000	2	0	2	1	1	1	1	3	3	1	2	2	2	2	2	3
Gabon	2005	20	9	21	2	13	17	21	27	4	4	4	2	4	4	4	4
Ghana	1999	5	4	6	1	4	3	5	6	3	3	3	2	4	2	3	2
Guinea-Bissau	2005	—	—	—	—	—	—	—	—	—	—	—	—	—	—	—	—
Kenya	1997	8	6	8	0	1	1	2	9	5	6	3	1	2	1	1	4
Madagascar	2001	2	2	2	0	0	1	2	2	1	1	0	0	0	1	1	0
Malawi	2003	10	7	11	1	1	3	4	12	14	12	9	3	1	6	6	10
Mauritania	2000	15	4	16	3	20	10	12	18	7	2	5	4	15	6	5	5
Morocco	2003	14	8	15	6	9	12	15	23	3	2	3	4	3	3	3	2
Mozambique	2003	13	6	14	3	6	6	9	15	20	12	12	15	17	13	16	10
Niger	2005	12	9	12	4	5	7	6	14	10	8	6	7	7	8	5	6

	Year																
Nigeria	2003	5	4	5	5	4	4	5	5	6	5	5	15	9	6	5	4
Rwanda	1998	10	5	11	—	—	1	4	11	10	7	3	—	—	1	4	4
São Tomé and Príncipe	2000	26	8	40	1	2	4	7	82	12	5	16	1	2	3	4	19
Senegal	2001	9	6	10	4	5	6	8	11	4	4	3	4	4	4	3	3
Sierra Leone	2003	9	6	9	0	1	2	5	12	8	7	6	0	2	3	4	5
South Africa	2000	12	5	16	2	4	6	9	26	2	2	2	2	3	3	3	2
Tanzania	2000	—	—	—	—	—	—	—	—	—	—	—	—	—	—	—	—
Uganda	2002	6	5	6	1	2	2	4	9	7	7	5	5	4	4	4	4
Zambia	2002	5	4	5	1	1	2	4	7	5	5	4	3	2	3	4	4
Overall		9	5	10	3	4	5	6	14	6	5	4	5	5	4	4	4
Income group																	
Low income		9	5	10	2	4	4	5	13	7	5	5	5	5	4	4	5
Middle income		10	5	11	3	6	7	10	16	3	2	2	3	3	3	3	2
Urbanization																	
Low		7	6	8	3	3	4	5	9	6	6	4	6	4	4	4	4
Medium		7	4	8	1	2	3	5	9	9	6	6	5	6	5	7	5
High		11	5	13	3	6	7	8	19	4	3	4	4	4	3	3	4
Region																	
East		6	4	7	1	1	1	3	8	6	5	3	3	2	2	3	4
West		10	6	11	4	6	6	7	17	6	5	5	6	6	5	4	5
South		8	5	10	1	2	4	6	12	9	6	5	5	5	5	6	5
Central		9	4	10	2	5	7	9	12	3	2	3	3	2	2	3	2

Source: Banerjee and others 2008.
Note: Q = quintile; — = data not available.
a. Sample average.

Table A5.4 Kerosene Expenditure and Its Share in Household Budget

Country	Year	Expenditure budget (2002 $)								Share in household budget (%)							
		National[a]	Rural	Urban	Q1	Q2	Q3	Q4	Q5	National	Rural	Urban	Q1	Q2	Q3	Q4	Q5
Angola	2000	—	—	—	—	—	—	—	—	—	—	—	—	—	—	—	—
Benin	2002	2	2	2	2	2	2	2	2	2	3	2	5	4	3	2	2
Burkina Faso	2003	1	1	1	1	1	1	1	1	1	1	1	2	2	1	1	1
Burundi	1998	1	1	4	1	1	1	2	2	2	2	2	4	3	2	2	2
Cameroon	2004	—	—	—	—	—	—	—	—	—	—	—	—	—	—	—	—
Cape Verde	2001	2	1	2	1	1	1	2	2	1	2	1	2	2	1	1	1
Chad	2001	—	—	—	—	—	—	—	—	—	—	—	—	—	—	—	—
Congo, Dem. Rep.	2005	1	1	1	1	1	1	1	1	1	1	1	1	1	1	1	1
Congo, Rep.	2002	—	—	—	—	—	—	—	—	—	—	—	—	—	—	—	—
Côte d'Ivoire	2005	3	3	4	2	3	3	3	4	1	2	1	3	2	2	1	1
Ethiopia	2000	1	0	1	0	0	0	1	1	1	1	1	1	1	1	1	1
Gabon	2005	4	6	3	5	5	4	4	4	1	2	1	5	1	1	1	1
Ghana	1999	4	3	5	2	3	2	7	3	2	2	2	3	3	2	4	1
Guinea-Bissau	2005	—	—	—	—	—	—	—	—	—	—	—	—	—	—	—	—
Kenya	1997	2	1	4	1	1	2	2	3	1	1	2	2	2	2	2	1
Madagascar	2001	3	2	4	0	0	0	11	3	1	1	1	0	0	0	4	1
Malawi	2003	2	1	13	1	1	1	1	6	3	2	10	2	2	2	2	5
Mauritania	2000	1	2	1	0	1	2	2	1	1	1	0	0	1	1	2	0
Morocco	2003	3	3	2	3	3	4	4	5	1	1	0	2	1	1	1	0
Mozambique	2003	—	—	—	—	—	—	—	—	—	—	—	—	—	—	—	—

	Year															
Niger	2005	—	—	—	—	—	—	—	—	—	—	—	—	—	—	—
Nigeria	2003	3	3	4	2	3	3	4	3	4	4	6	4	4	4	3
Rwanda	1998	1	1	2	1	1	1	1	2	1	1	2	2	2	1	1
São Tomé and Príncipe	2000	—	—	—	—	—	—	—	—	—	—	—	—	—	—	—
Senegal	2001	1	1	1	1	1	1	1	1	0	1	1	1	1	1	0
Sierra Leone	2003	3	2	3	2	2	3	4	3	2	3	4	3	3	3	1
South Africa	2000	0	0	0	0	0	0	1	0	0	0	0	0	0	0	0
Tanzania	2000	—	—	—	—	—	—	—	—	—	—	—	—	—	—	—
Uganda	2002	1	1	1	1	1	1	2	2	1	2	2	2	2	2	1
Zambia	2002	—	—	—	—	—	—	—	—	—	—	—	—	—	—	—
Overall		2	2	3	1	2	2	3	3	2	2	2	2	2	2	1
Income group																
Low income		2	2	3	1	2	3	0	2	2	2	3	2	2	2	0
Middle income		2	3	2	2	2	2	3	3	1	1	2	1	1	1	1
Urbanization																
Low		2	1	4	1	1	1	3	3	2	2	2	2	2	2	1
Medium		2	2	2	2	2	2	0	2	2	2	3	3	2	2	0
High		2	3	2	2	2	2	3	3	1	1	2	1	1	2	1
Region																
East		1	1	2	1	1	1	2	2	1	1	2	2	2	2	1
West		2	2	3	2	2	3	3	3	2	1	3	2	2	2	1
South		2	1	6	0	0	0	4	4	2	4	1	1	1	2	2
Central		3	4	2	3	3	2	3	3	1	1	3	1	1	1	1

Source: Banerjee and others 2008.
Note: Q = quintile; — = data not available.
a. Sample average.

Table A5.5 Liquefied Propane Gasoline (LPG) Expenditure and Its Share in Household Budget

Country	Year	Expenditure budget (2002 $)								Share in household budget (%)							
		National[a]	Rural	Urban	Q1	Q2	Q3	Q4	Q5	National	Rural	Urban	Q1	Q2	Q3	Q4	Q5
Angola	2000	—	—	—	—	—	—	—	—	—	—	—	—	—	—	—	—
Benin	2002	1	1	1	1	1	1	1	1	1	1	1	1	1	1	1	0
Burkina Faso	2003	5	5	5	5	5	5	5	5	4	5	2	10	7	5	4	2
Burundi	1998	—	—	—	—	—	—	—	—	—	—	—	—	—	—	—	—
Cameroon	2004	—	—	—	—	—	—	—	—	—	—	—	—	—	—	—	—
Cape Verde	2001	3	2	3	1	2	3	3	3	2	3	2	2	3	3	3	2
Chad	2001	13	12	13	7	9	11	13	15	4	6	3	7	6	5	4	2
Congo, Dem. Rep.	2005	1	0	2	1	0	0	0	1	1	0	1	2	0	0	0	1
Congo, Rep.	2002	—	—	—	—	—	—	—	—	—	—	—	—	—	—	—	—
Côte d'Ivoire	2005	5	4	5	2	3	4	5	5	2	3	2	2	2	2	2	1
Ethiopia	2000	1	0	1	1	0	0	0	2	1	0	2	3	1	0	0	2
Gabon	2005	11	11	11	9	11	11	11	10	2	4	2	10	3	3	2	2
Ghana	1999	6	5	6	—	—	3	6	6	4	4	3	—	—	2	3	2
Guinea-Bissau	2005	—	—	—	—	—	—	—	—	—	—	—	—	—	—	—	—
Kenya	1997	14	11	14	—	—	2	9	14	10	10	6	—	—	2	7	6
Madagascar	2001	6	6	6	—	8	—	14	5	2	3	1	—	6	—	5	1
Malawi	2003	—	—	—	—	—	—	—	—	—	—	—	—	—	—	—	—
Mauritania	2000	2	1	4	0	1	2	3	4	1	1	1	0	1	1	1	1
Morocco	2003	10	9	11	6	8	10	13	16	2	3	2	4	3	3	2	2
Mozambique	2003	—	—	—	—	—	—	—	—	—	—	—	—	—	—	—	—
Niger	2005	8	2	10	1	1	1	2	10	6	2	5	2	1	1	2	4

Nigeria	2003	0	0	0	0	0	0	0	0	0	0	0	0	0	0	0	0
Rwanda	1998	3	1	20	0	1	1	7	3	1	6	1	1	1	6	1	3
São Tomé and Príncipe	2000	—	—	—	—	—	—	—	—	—	—	—	—	—	—	—	—
Senegal	2001	6	3	7	2	3	4	8	3	2	2	2	2	2	2	3	0
Sierra Leone	2003	5	1	6	0	5	3	7	5	1	4	1	4	4	2	2	3
South Africa	2000	0	0	0	0	0	0	0	0	0	0	0	0	0	0	0	0
Tanzania	2000	—	—	—	—	—	—	—	—	—	—	—	—	—	—	—	—
Uganda	2002	—	—	—	—	—	—	—	—	—	—	—	—	—	—	—	—
Zambia	2002	6	3	10	1	1	1	5	12	6	3	8	2	2	2	5	7
Overall		5	4	7	2	3	3	5	7	3	2	3	3	2	2	2	2
Income group																	
Low income		5	3	7	2	3	3	4	6	3	3	3	3	2	2	3	2
Middle income		6	6	6	4	5	6	7	7	2	2	1	4	2	2	2	1
Urbanization																	
Low		7	5	10	3	4	3	6	8	4	4	4	5	4	3	3	3
Medium		3	1	5	1	1	1	2	5	3	1	3	2	1	2	2	3
High		5	4	5	3	4	4	5	6	2	2	2	2	2	2	2	1
Region																	
East		6	4	12	1	0	3	3	7	5	4	5	2	4	1	1	3
West		5	3	5	2	3	4	4	6	3	2	2	2	2	2	2	2
South		4	3	5	0	3	1	6	6	3	2	3	1	3	3	3	3
Central		8	8	8	6	7	7	8	9	2	3	2	6	3	3	2	2

Source: Banerjee and others 2008.
Note: Q = quintile; — = data not available.
a. Sample average.

Table A5.6 Wood/Charcoal Expenditure and Its Share in Household Budget

Country	Year	Expenditure budget (2002 $)								Share in household budget (%)							
		National[a]	Rural	Urban	Q1	Q2	Q3	Q4	Q5	National	Rural	Urban	Q1	Q2	Q3	Q4	Q5
Angola	2000	—	—	—	—	—	—	—	—	—	—	—	—	—	—	—	—
Benin	2002	4	4	4	4	4	3	4	4	4	5	3	9	6	5	4	3
Burkina Faso	2003	4	4	4	4	4	4	4	4	3	4	2	8	5	4	3	2
Burundi	1998	8	4	11	2	3	3	4	9	12	6	5	9	8	6	7	8
Cameroon	2004	3	3	3	0	1	1	2	5	3	3	2	1	1	2	2	3
Cape Verde	2001	3	3	3	3	3	3	3	3	3	4	2	5	4	3	3	2
Chad	2001	6	8	5	4	4	5	9	9	2	4	1	4	2	2	3	1
Congo, Dem. Rep.	2005	1	1	2	1	1	1	1	2	1	1	1	1	1	1	1	1
Congo, Rep.	2002	—	—	—	—	—	—	—	—	—	—	—	—	—	—	—	—
Côte d'Ivoire	2005	5	6	5	4	5	6	5	5	2	3	2	5	4	3	2	1
Ethiopia	2000	4	4	2	3	4	4	4	5	7	8	3	9	8	8	7	6
Gabon	2005	5	6	5	4	5	4	5	6	1	2	1	4	2	1	1	1
Ghana	1999	6	4	6	2	4	4	8	6	3	3	3	3	3	3	4	2
Guinea-Bissau	2005	—	—	—	—	—	—	—	—	—	—	—	—	—	—	—	—
Kenya	1997	4	4	4	2	3	3	4	4	3	3	2	4	4	3	3	2
Madagascar	2001	2	2	2	3	1	1	2	2	1	1	0	3	1	1	1	0
Malawi	2003	6	6	8	5	6	6	7	8	9	10	6	14	13	12	10	7
Mauritania	2000	3	2	4	1	2	3	4	3	1	1	1	1	2	2	1	1
Morocco	2003	2	3	2	2	2	2	3	3	1	1	0	1	1	1	1	0
Mozambique	2003	0	0	0	0	0	0	0	1	1	1	0	1	1	1	1	0
Niger	2005	6	5	7	4	4	5	6	7	5	5	4	8	5	5	5	3
Nigeria	2003	1	1	2	1	2	2	1	1	2	2	2	5	3	2	2	1

	Year																
Rwanda	1998	8	5	10	1	1	3	4	11	8	7	3	3	3	4	4	4
São Tomé and Príncipe	2000	—	—	—	—	—	—	—	—	—	—	—	—	—	—	—	—
Senegal	2001	3	1	1	0	1	0	1	1	0	0	0	0	1	0	0	0
Sierra Leone	2003	3	2	4	1	2	2	3	4	3	3	2	3	3	3	3	2
South Africa	2000	3	4	3	3	4	5	3	2	1	2	0	3	3	2	1	0
Tanzania	2000	—	—	—	—	—	—	—	—	—	—	—	—	—	—	—	—
Uganda	2002	4	3	4	2	2	3	4	6	5	5	3	7	5	5	4	3
Zambia	2002	4	2	7	1	1	1	2	9	4	2	5	3	3	2	2	5
Overall		4	3	4	2	3	3	4	5	3	3	2	5	4	3	3	2
Income group																	
Low income		4	3	5	2	3	3	4	5	4	4	2	5	4	4	3	3
Middle income		3	4	3	2	3	3	3	4	2	2	1	3	2	2	1	1
Urbanization																	
Low		5	5	6	3	3	4	5	6	5	5	3	7	5	5	5	4
Medium		3	2	3	1	2	2	2	4	3	2	2	3	3	2	2	2
High		3	3	3	2	3	3	3	4	2	2	1	3	2	2	2	1
Region																	
East		5	4	6	2	3	3	4	7	7	6	3	7	6	5	5	4
West		3	3	4	2	3	3	4	4	2	3	2	4	3	3	3	2
South		3	3	4	2	3	3	3	4	3	3	2	5	4	3	3	2
Central		4	5	4	2	3	3	4	6	2	3	1	3	2	2	2	2

Source: Banerjee and others 2008.

Note: Q = quintile; — = data not available.

a. Sample average.

Table A5.7 Rural Access to Power, Off-Grid Power, and Rural Electrification Agency and Fund

	Residential access, rural (%)	Rural population with access, number of people	Estimated rural population served by off-grid power, number of people	Estimated rural population with access served by off-grid power (%)	Existence of rural electrification agency	Existence of rural electrification fund
Benin	6	269,301	93,000	34.53	Yes	Yes
Burkina Faso	1	84,560	36,250	42.87	Yes	Yes
Cameroon	16	1,209,150			Yes	No
Central African Republic	1	12,838				
Chad	0	21,745			No	No
Comoros	21	79,027				
Congo, Rep.	16	250,773				
Côte d'Ivoire	27	2,615,106	500,000	19.12	No	Yes
Ethiopia	2	1,145,293	192,500	16.81	Yes	Yes
Gabon	31	74,774				
Ghana	21	2,378,302	25,000	1.05	No	No
Guinea	3	195,013				
Kenya	4	928,828	397,500	42.80	No	Yes

Lesotho	1	12,370			No	Yes
Madagascar	10	1,258,024			Yes	Yes
Malawi	2	256,185			No	Yes
Mali	3	238,364				
Mauritania	3	47,275			Yes	Yes
Mozambique	1	189,700			No	Yes
Namibia	10	135,965			No	No
Niger	0	28,564			Yes	Yes
Nigeria	35	23,286,973	4.56	1,062,500	No	No
Rwanda	1	107,879	14.83	16,000	Yes	Yes
Senegal	19	1,240,233	22.98	285,000	No	Yes
South Africa	36	6,824,964			Yes	Yes
Tanzania	2	498,401				
Togo	2	71,490				
Uganda	3	610,555	75.75	462,500	Yes	Yes
Zambia	3	254,274			Yes	Yes
Zimbabwe	7	617,540				

Source: Banerjee and others 2008.
Note: Blank cells = data not available.

Table A5.8   Share of Urban Households Whose Utility Bill Would Exceed 5 Percent of the Monthly Household Budget at Various Prices
percent

Group	Country	\$2	\$4	\$6	\$8	\$10	\$12	\$14	\$16
					Monthly bill				
1	Cape Verde	0	0	0	0	0	0	0	0
	Morocco	0	0	0	0	0	0	0	0
	Senegal	0	0	0	0	0	0	1	1
	South Africa	0	0	0	0	1	1	1	1
	Cameroon	0	0	0	0	1	2	7	17
	Côte d'Ivoire	0	0	1	2	3	5	7	10
	Congo, Rep.	0	0	3	5	12	21	28	35
2	Ghana	0	2	7	11	30	46	55	67
	Benin	0	2	4	12	33	45	60	71
	Kenya	0	0	5	20	36	62	72	78
	Sierra Leone	0	4	16	30	44	54	62	67
	São Tomé and Príncipe	0	2	13	29	46	64	77	81
	Burkina Faso	0	4	20	34	47	62	72	78
	Zambia	0	4	18	35	50	58	67	76
	Nigeria	3	10	23	35	57	78	89	95
	Madagascar	0	16	28	47	61	68	78	85
3	Niger	1	11	28	55	70	79	89	93
	Tanzania	1	8	25	55	75	89	96	98
	Guinea-Bissau	0	6	38	65	81	89	91	93
	Uganda	2	17	45	65	82	90	96	97
	Burundi	7	29	53	72	82	90	97	100
	Malawi	2	32	66	78	87	92	93	94
	Congo, Dem. Rep.	9	49	79	91	98	99	100	100
	Ethiopia	40	87	95	99	99	99	99	100
Summary	Low income	5.0	18.4	32.4	44.5	59.5	72.3	79.7	84.3
	Middle income	0.0	0.0	0.1	0.2	1.2	1.8	2.9	4.7
	All	3.7	13.7	24.2	33.2	44.7	54.3	60.2	64.1

Source: Banerjee and others 2008.

Table A5.9   Overall Targeting Performance (Ω) of Utility Subsidies

Country	Omega (Ω) value
Burkina Faso	0.06
Burundi	0.10
Cameroon	0.36
Cape Verde	0.48
Central African Republic	0.27
Chad	0.06
Congo, Rep.	0.62
Côte d'Ivoire	0.51
Gabon	0.78
Ghana	0.31
Guinea	0.22
Mozambique	0.31
Nigeria	0.79
Rwanda	0.01
São Tomé and Príncipe	0.41
Senegal	0.41
Togo	0.47
Uganda	0.02

Source: Banerjee and others 2008.

Table A5.10   Potential Targeting Performance of Connection Subsidies under Different Subsidy Scenarios

Country	Scenario 1 *Distribution of connection subsidies mirrors distribution of existing connections*	Scenario 2 *Only households with access but no connection receive subsidy*	Scenario 3 *All unconnected households receive subsidy*
Burkina Faso	0.08	0.64	1.10
Burundi	0.23	0.83	1.03
Cameroon	0.46	1.17	1.40
Cape Verde	0.55	1.27	1.35
Central African Republic	0.36	0.73	1.02
Chad	0.12	0.58	1.01
Congo, Rep.	0.41	1.02	1.23
Côte d'Ivoire	0.61	1.33	1.33
Gabon	0.75	1.17	1.30
Ghana	0.38	0.98	1.52
Guinea	0.25	0.52	1.15
Mozambique	0.35	1.08	1.06
Nigeria	0.77	1.09	1.10
Rwanda	0.03	0.47	1.05
São Tomé and Príncipe	0.56	1.15	1.33
Senegal	0.63	1.23	1.22
Togo	0.47	0.92	1.18
Uganda	0.06	0.87	1.08

Source: Banerjee and others 2008.

Table A5.11 Value of Cost Recovery Bill at Consumption of 50 kWh/Month

Country	Based on total historical cost							Based on LRMC						
			% of monthly budget							% of monthly budget				
	$/month	National[a]	Q1	Q2	Q3	Q4	Q5	$/month	National[a]	Q1	Q2	Q3	Q4	Q5
Benin	10	11	24	17	14	12	7	10	11	23	16	13	11	7
Burkina Faso	8	6	16	11	9	7	3	13	10	26	19	15	11	6
Cameroon	9	8	18	12	10	7	5	4	3	7	5	4	3	2
Cape Verde	9	7	14	11	9	8	5	—	—	—	—	—	—	—
Chad	7	2	7	4	3	2	1	4	1	4	2	2	1	1
Congo, Dem. Rep.	3	3	8	5	4	3	2	2	2	4	3	2	2	1
Congo, Rep.	10	5	14	9	7	5	3	3	1	4	3	2	2	1
Côte d'Ivoire	5	2	7	4	3	2	1	8	3	9	5	4	3	2
Ethiopia	4	7	13	10	8	7	5	10	17	30	23	19	16	11
Ghana	6	4	10	6	4	3	2	5	3	8	5	4	3	2
Kenya	7	5	13	9	7	5	3	6	4	11	7	6	4	3
Madagascar	7	3	8	5	4	3	1	—	—	—	—	—	—	—
Malawi	5	6	14	10	8	7	4	3	4	8	6	5	4	2
Mali	17	26	73	51	39	31	12	13	19	54	38	29	23	9
Niger	16	13	33	23	18	14	7	13	10	26	18	14	11	5
Nigeria	5	6	16	9	7	5	3	7	8	22	13	9	7	5
Rwanda	8	8	28	17	13	9	3	6	6	20	12	9	7	2
Senegal	6	3	6	4	3	3	1	22	9	21	16	13	10	5
South Africa	3	1	4	2	2	1	0	3	1	4	2	2	1	0
Tanzania	7	12	25	18	14	11	8	5	8	18	13	10	8	5
Uganda	5	6	20	11	8	5	2	6	8	23	13	9	6	3
Zambia	3	3	9	6	4	3	2	4	4	11	7	5	4	2

Source: Briceño-Garmendia and Shkaratan 2010.

Note: kWh = kilowatt-hour; LRMC = long-run marginal cost; Q = quintile. — = data not available.

a. Sample average.

Table A5.12   Residential Tariff Schedules

Country	Tariff type	Fixed charge/ month Yes/no	Number of blocks	Border between first and second block range (kWh)	Range of block prices (cents/kWh)
Benin	IBT	No	3	20	9.6–16.3
Botswana	FR	Yes	1	n.a.	5.9
Burkina Faso	IBT	Yes	3	50	18.4–20.8
Cameroon	IBT	No	3	50	8.6–12.0
Cape Verde	IBT	No	2	40	22.5–28.0
Chad	IBT	No	3	30	15.7–38.1
Congo, Dem. Rep.	IBT	No	11	100	3.98–8.52
Congo, Rep.	FR	Yes	1	n.a.	11.0
Côte d'Ivoire	IBT	Yes	2	40	6.9–14.2
Ethiopia	IBT	Yes	7	50	3.2–8.0
Ghana	IBT	Yes	3	300	7.6–15.3
Kenya	IBT	Yes	4	50	4.9–44.0
Lesotho	FR	No	1	n.a.	7.2
Madagascar	FR	Yes	1	n.a.	7.6
Malawi[a]	IBT/FR	Yes/no	3/1	30	2.0–4.1/3.1
Mali	IBT	No	4	200	26.6–31.0
Mozambique[a]	IBT/FR	Yes/no	4/1	100	4.0–12.1/11.0
Namibia	FR	No	1	n.a.	11.7
Niger	FR	Yes	1	n.a.	13.6
Nigeria	IBT	Yes	5	20	0.9–6.5
Rwanda	FR	No	1	n.a.	14.6
Senegal[a]	IBT	No	3	150	23.8–26.2
South Africa	IBT	No	2	50	0.0–7.2
Sudan	—	—	—	—	—
Tanzania[a]	IBT/FR	No/yes	2/1	50	4.1–13.0/10.8
Uganda	IBT	Yes	2	15	3.4–23.3
Zambia	IBT	Yes	3	300	1.6–3.7
Zimbabwe	IBT	No	3	50	0.6–13.5

Source: Briceño-Garmendia and Shkaratan 2010.
Note: FR = fixed rate; IBT = increasing block tariff; kWh = kilowatt-hour; n.a. = not applicable; — = data not available.
a. The country has two tariffs, equally applicable, for typical residential customers.

Table A5.13   Social Tariff Schedules

Country	Type of tariff	Fixed charge ($/month)	Block border	Price per block (cents/kWh)
Benin	Social tranche	n.a.		9.6
Botswana	n.a.	n.a.		n.a.
Burkina Faso	Block 1	0.18		14.3
Cameroon[a]	Block 1 residential	12.90		8.6
Cape Verde	Block 1, residential	—		22.5
Chad	Block 1 residential	n.a.		15.7
Congo, Dem. Rep.	Social tariff	0.01		4.0
Congo, Rep.	n.a.	n.a.		n.a.
Côte d'Ivoire	Block 1 residential	0.64		6.9
Ethiopia	Block 1 residential	0.16		3.2
Ghana	Block 1 residential	0.54		7.6
Kenya	Block 1 residential	1.74		4.9
Lesotho	—	—		—
Madagascar	Economic tariff	0.30	25	6.0
			>25	27.6
Malawi	Block 1 residential	0.92		2.0
Mali	Social tariff	n.a.	50	13.2
			100	20.3
			200	23.9
			>200	27.7
Mozambique	Block 1 residential	n.a.		4.0
Namibia	n.a.	n.a.		n.a.
Niger	—	—		—
Nigeria	Pensioners' tariff	0.23		3.0
Rwanda	—	—		—
Senegal	Tranche 1 residential	n.a.	150	0.24
South Africa	Block 1 residential	n.a.		—
Sudan	—	—		—
Tanzania	n.a.	n.a.		3.0
Uganda	Block 1 residential	1.09		3.4
Zambia	Block 1 residential	1.31		1.6
Zimbabwe	Tranche 1 residential	n.a.		0.6

Source: Briceño-Garmendia and Shkaratan 2010.
Note: kWh = kilowatt-hour; n.a. = not applicable; — = data not available.
a. In Cameroon fixed residential charge is 2,500 per kW if subscribed load is up to 200 hours and 4,200 per kW if it is above 200 hours.

Table A5.14   Industrial Tariff Schedules

Country	Tariff type	Fixed charge/month Yes/no	Demand charge Yes/no	Number of blocks	Range of block prices (cents/kWh)
Benin	FR	No	No	1	15.1
Botswana	FR	No	No	1	6.7
Burkina Faso	TOU	Yes	Yes	2	31.6–16.8
Cameroon	DBT	No	Yes	2	11.3–9.9
Cape Verde	FR	No	No	1	21.8
Chad	IBT	No	Yes	3	15.9–40.0
Congo, Dem. Rep.	DBT	No	No	5	11.1–10.7
Congo, Rep.	FR	Yes	No	1	9.7
Côte d'Ivoire	DBT	Yes	No	2	18.6–15.9
Ethiopia	TOU	Yes	No	3	6.7–6.3
Ghana	IBT	Yes	No	3	11.1–16.0
Kenya	FR	Yes	No	1	21.4
Lesotho	FR	No	Yes	1	1.2
Madagascar	FR	Yes	Yes	1	16.9
Malawi	FR	Yes	Yes	1	3.0
Mali	FR	No	No	1	23.2
Mozambique	FR	Yes	Yes	1	5.4
Namibia	FR	Yes	Yes	1	8.4
Niger	FR	Yes	Yes	1	12.2
Nigeria	IBT	Yes	No	4	5.0–6.5
Rwanda	FR	No	No	1	17.2
Senegal	TOU	Yes	No	2	14.4–20.8
South Africa	IBT/FR	Yes	No	3/1	4.0–9.5
Tanzania	FR	Yes	Yes	1	5.3
Uganda	TOU	Yes	No	1	21.8
Zambia	FR	Yes	No	1	3.7

Source: Briceño-Garmendia and Shkaratan 2010.
Note: DBT = decreasing block tariff; FR = fixed rate; IBT = increasing block tariff; TOU = time of use; kWh = kilowatt-hour.

Table A5.15 Commercial Tariff Schedules

Country	Tariff type	Fixed charge/month Yes/no	Demand charge Yes/no	Number of blocks	Range of block prices (cents/kWh)
Benin	FR	No	No	1	10.7
Botswana	FR	Yes	Yes	1	3.1
Burkina Faso	TOU	Yes	Yes	2	22.6–10.3
Cameroon	TOU	No	Yes	2	8.7–8.5
Cape Verde	FR	No	No	1	17.7
Chad	TOU	No	Yes	3	20.5–37.9
Congo, Dem. Rep.	DBT	No	No	5	15.2–14.6
Congo, Rep.	FR	Yes	Yes	1	11.2
Côte d'Ivoire	TOU	Yes	No	3	10.7–8.8
Ethiopia	TOU	Yes	No	3	4.7–5.9
Ghana	FR	Yes	Yes	1	5.4
Kenya	DBT	Yes	Yes	3	16.4–14.0
Lesotho	FR	No	Yes	1	1.1
Madagascar	FR	Yes	Yes	1	9.9
Malawi	FR	Yes	Yes	1	2.4
Mali	—	—	—	—	16.9
Mozambique	FR	Yes	Yes	1	4.5
Namibia	FR	Yes	Yes	1	12.4
Niger	FR	Yes	Yes	1	8.8
Nigeria	IBT	Yes	Yes	5	5.0–6.5
Rwanda	FR	No	No	1	17.2
Senegal	TOU	Yes	No	2	13.0–18.7
South Africa	TOU	Yes	Yes	2	2.6–1.8
Tanzania	FR	Yes	Yes	1	4.9
Uganda	TOU	Yes	Yes	1	16.7
Zambia	DBT	Yes	Yes	4	2.2–1.2

Source: Briceño-Garmendia and Shkaratan 2010.
Note: DBT = decreasing block tariff; FR = fixed rate; IBT = increasing block tariff; TOU = time of use; kWh = kilowatt-hour. — = data not available.

Table A5.16   Value and Volume of Sales to Residential Customers as Percentage of Total Sales

Country	Value of sales (%)	Volume of sales (%)
Benin	—	96
Burkina Faso	63	63
Cameroon	60	33
Cape Verde	56	56
Chad	67	63
Congo, Dem. Rep.	47	70
Côte d'Ivoire	47	—
Ethiopia	27	56
Ghana	35	71
Kenya	42	39
Lesotho	—	30
Madagascar	—	61
Malawi	—	36
Mozambique	70	59
Namibia	36	47
Niger	59	—
Nigeria	39	51
Rwanda	50	—
Senegal	63	59
South Africa	17	51
Tanzania	48	44
Uganda	—	36

Source: Briceño-Garmendia and Shkaratan 2010.
Note: — = data not available.

Recommitting to the Reform of State-Owned Enterprises

Table A6.1 Electricity Sector Tariffs and Costs

cents/kWh

	Effective tariffs			Costs			
	Residential at 100 kWh/month	Commercial at 900 kWh/month	Industrial at demand level of 100 kVA	Historical operating costs	Historical total costs	Average revenue	LRMC
Benin	13.6	15.1	10.7	11.6	19.8	14.2	19.0
Botswana	7.5	7.2	4.0	11.9	13.9	18.4	6.0
Burkina Faso	20.0	26.7	15.0	4.4	15.1	19.7	25.0
Cameroon	10.9	11.4	9.2	12.7	17.1	10.9	7.0
Cape Verde	25.8	21.8	17.7	14.3	17.9	18.0	—
Chad	30.0	44.7	38.8	9.4	13.7	32.1	7.0
Congo, Dem. Rep.	4.0	11.0	14.6	3.9	6.8	4.3	4.0
Congo, Rep.	16.0	10.7	11.2	13.4	20.1	12.8	6.0
Côte d'Ivoire	11.9	16.9	10.7	6.6	10.9	1.0	15.0
Ethiopia	4.1	8.3	4.7	2.1	8.5	6.0	19.0
Ghana	8.2	13.9	6.4	7.5	12.4	8.0	10.0
Kenya	14.8	21.7	15.1	8.4	14.2	14.0	12.0
Lesotho	7.2	9.3	3.3	6.4	10.8	7.1	6.0

Madagascar	3.0	25.3	10.5	10.5	15.0	45.9	
Malawi	4.0	6.9	3.1	5.9	9.1	3.2	5.0
Mali	26.6	23.2	—	16.3	33.6	18.6	25.0
Mozambique	6.8	8.0	5.1	6.3	9.0	7.6	4.0
Namibia	11.7	14.0	13.6	7.3	11.3	12.4	11.0
Niger	14.1	13.2	9.3	23.4	32.1	15.5	25.0
Nigeria	3.4	5.0	5.1	2.2	9.7	2.8	13.0
Rwanda	14.6	17.2	17.2	6.8	16.6	22.2	12.0
Senegal	23.8	22.8	15.8	19.4	25.0	14.9	43.0
South Africa	3.6	7.7	2.7	3.4	6.0	16.0	6.0
Tanzania	6.7	8.0	5.4	8.0	14.1	7.5	10.0
Uganda	21.4	21.9	17.0	5.3	10.4	8.7	12.0
Zambia	2.9	4.4	2.5	3.6	6.5	5.0	8.0

Source: Briceño-Garmendia and Shkaratan 2010.

Note: kVA = kilovolt-ampere; kWh = kilowatt-hour; LRMC = long-run marginal cost. Effective tariffs are prices per kWh at typical monthly consumption levels calculated using tariff schedules that are applicable to typical customers within each customer group. — = data not available.

Table A6.2 Residential Effective Tariffs at Different Consumption Level

cents/kWh

	50 kWh	75 kWh	100 kWh	150 kWh	200 kWh	300 kWh	400 kWh	450 kWh	500 kWh	900 kWh
Benin	12.6	13.3	13.6	14.0	14.1	19.7	22.5	23.5	24.2	27.2
Botswana	9.1	8.0	7.5	6.9	6.7	6.4	6.3	6.2	6.2	6.0
Burkina Faso	20.6	20.2	20.0	19.9	19.8	20.1	20.3	20.4	20.4	20.6
Cameroon	8.6	10.9	10.9	10.9	10.9	12.0	12.0	12.0	12.0	12.0
Cape Verde	23.6	25.1	25.8	26.5	26.9	27.3	27.4	27.5	27.5	27.7
Chad	22.9	27.3	30.0	32.7	34.1	35.4	36.1	36.3	36.5	37.2
Congo, Dem. Rep.	4.0	4.0	4.0	4.0	4.0	3.9	3.9	3.9	3.9	5.5
Congo, Rep.	21.1	17.7	16.0	14.3	13.5	12.6	12.2	12.1	12.0	11.5
Côte d'Ivoire	9.6	11.1	11.9	12.6	13.0	13.4	13.6	13.6	13.7	13.9
Ethiopia	3.9	4.1	4.1	5.3	5.6	6.1	6.2	6.4	6.6	7.2
Ghana	8.7	8.4	8.2	8.0	7.9	7.8	9.1	9.6	9.9	11.8
Kenya	8.4	12.7	14.8	16.9	18.0	19.1	19.9	20.1	20.4	21.2
Lesotho	7.2	7.2	7.2	7.2	7.2	7.2	7.2	7.2	7.2	7.2
Madagascar	6.0	4.0	3.0	2.0	1.5	1.0	0.7	0.7	0.6	0.3
Malawi	4.8	4.3	4.0	3.8	3.7	3.6	3.5	3.5	3.5	3.4
Mali	26.6	26.6	26.6	26.6	26.6	28.1	28.8	29.1	29.3	30.0
Mozambique	9.6	7.7	6.8	7.4	7.7	9.0	9.6	9.8	10.0	10.9
Namibia	11.7	11.7	11.7	11.7	11.7	11.7	11.7	11.7	11.7	11.7
Niger	14.5	14.2	14.1	13.9	13.9	13.8	13.7	13.7	13.7	13.7
Nigeria	2.5	3.8	3.4	3.8	4.2	4.9	5.3	5.4	5.6	6.0
Rwanda	14.6	14.6	14.6	14.6	14.6	14.6	14.6	14.6	14.6	14.6
Senegal	23.8	23.8	23.8	23.8	24.2	24.8	25.1	25.2	25.3	25.7
South Africa	0.0	2.4	3.6	4.8	5.4	6.0	6.3	6.4	6.5	6.8
Tanzania	3.2	5.5	6.7	7.9	8.5	9.0	8.8	8.8	8.8	8.6
Uganda	19.5	20.7	21.4	22.0	22.3	22.6	22.8	22.8	22.9	23.1
Zambia	4.2	3.3	2.9	2.4	2.2	2.0	2.1	2.1	2.1	2.5
Zimbabwe	0.6	3.0	4.3	5.5	6.1	6.7	7.0	7.1	7.2	10.0

Source: Briceño-Garmendia and Shkaratan 2010.

Note: kWh = kilowatt-hour. See note to table A6.1 regarding effective tariffs.

Table A6.3   Electricity Sector Efficiency

	System losses (% production)	Connections per sector employee	Collection ratio (implicit, revenue to tariff)	Cost recovery (%, ratio of residential effective tariff to total historical cost)	Residential effective tariff/LRMC (%)
Benin	18	148	100	69	72
Botswana	10		62	54	125
Burkina Faso	25	179	88	133	80
Cameroon	31	180	106	64	156
Cape Verde	17	112	77	144	
Chad	33	43	91	220	429
Congo, Dem. Rep.	40	—	107	59	100
Congo, Rep.	47	—	83	80	267
Côte d'Ivoire	—	57	7	109	79
Ethiopia	22	84	108	48	22
Ghana	25	146	90	66	82
Kenya	18	227	85	104	123
Lesotho	20	95	108	67	121
Madagascar	24	—	397	20	—
Malawi	23	—	79	44	81
Mali	22	—	72	79	106
Mozambique	25	99	102	75	169
Namibia	15	38	107	103	106
Niger	27	118	110	44	56
Nigeria	30	127	67	35	26
Rwanda	23	189	152	88	121
Senegal	21	257	66	95	55
South Africa	10	132	277	60	60
Tanzania	26	124	117	48	67
Uganda	36	444	76	206	178
Zambia	12	—	173	44	36

Source: Eberhard and others 2008.
Note: LRMC = long-run marginal cost. — = data not available.

Table A6.4 Hidden Costs of Power Utilities as a Percentage of GDP and Utility Revenue

Percent

	Percent of revenues				Percent of GDP			
	T&D losses	Underpricing	Undercollection of bills	Overstaffing	T&D losses	Underpricing	Undercollection of bills	Overstaffing
Benin	12.8	39.1	0.5	13.8	0.2	0.7	0.0	0.2
Botswana	0.7	138.7	61.1	—	0.0	1.8	0.8	—
Burkina Faso	12.5	0.0	14.7	9.5	0.2	0.0	0.3	0.2
Cameroon	36.3	57.9	0.0	8.3	0.8	1.2	0.0	0.2
Cape Verde	8.1	0.0	29.6	20.8	0.2	0.0	0.9	0.6
Chad	11.0	0.0	9.1	23.6	0.0	0.0	0.0	0.1
Congo, Dem. Rep.	163.6	201.6	0.0	—	1.3	1.6	0.0	—
Congo, Rep.	63.1	30.9	21.0	30.9	0.6	0.3	0.2	0.3
Côte d'Ivoire	—	0.0	417.1	24.2	—	0.0	4.4	0.3
Ethiopia	18.6	33.5	6.3	—	0.2	0.3	0.1	—
Ghana	26.5	52.4	2.1	—	0.7	1.5	0.1	—
Kenya	9.1	0.0	34.6	5.1	0.3	0.0	1.1	0.2
Lesotho	16.9	32.5	19.5	—	0.3	0.6	0.3	—

Madagascar	5.0	2.3	0.0	—	0.3	0.2	0.0	—
Malawi	40.5	105.3	75.1	—	0.5	1.3	0.9	—
Mali	23.4	36.8	39.1	6.4	0.6	1.0	1.0	0.2
Mozambique	19.9	15.0	4.6	17.7	0.3	0.2	0.1	0.3
Namibia	51.6	0.0	—	—	0.1	0.0	—	—
Niger	39.1	116.5	0.0	12.5	0.6	1.8	0.0	0.2
Nigeria	76.8	195.1	50.3	—	0.4	1.0	0.3	—
Rwanda	10.8	9.3	0.0	6.8	0.2	0.1	0.0	0.1
Senegal	9.6	0.0	10.8	5.4	0.3	0.0	0.3	0.2
South Africa	0.0	5.9	0.0	—	0.0	1.0	0.0	—
Tanzania	33.5	90.9	0.0	6.1	0.5	1.3	0.0	0.1
Uganda	34.6	0.0	39.4	5.2	0.6	0.0	0.7	0.1
Zambia	2.9	72.9	2.3	—	0.0	1.2	0.0	—

Source: Eberhard and others 2008.

Note: Unaccounted losses = end-user consumption x average cost recovery price x (total loss rate – normative loss rate) / (1 – normative loss rate). Underpricing = end-user consumption x (average cost recovery price – average actual tariff). Collection inefficiencies = end-user consumption x average actual tariff x (1 – collection rate). GDP = gross domestic product; T&D = transmission and distribution. — = data not available.

Closing Africa's Power Funding Gap

Table A7.1 Existing Spending on the Power Sector[a]

| | GDP share (%) | | | | | | | Current US$, million p.a. | | | | | | |
| | O&M | Capital expenditure | | | | | | O&M | Capital expenditure | | | | | |
Country	Public sector	Public sector	ODA	Non-OECD financiers	PPI	Total CAPEX	Total spending	Public sector	Public sector	ODA	Non-OECD financiers	PPI	Total CAPEX	Total spending
Benin	1.65	0.68	0.31	0.01	0.00	1.01	2.66	71	29	13	1	0	43	114
Botswana	0.97	0.33	0.01	0.00	0.00	0.33	1.30	101	35	1	0	0	35	137
Burkina Faso	0.96	0.19	0.61	0.01	0.00	0.81	1.76	52	10	33	1	0	44	96
Cameroon	1.04	0.05	0.09	0.01	0.36	0.51	1.55	173	8	15	2	60	84	258
Cape Verde	3.24	2.30	0.01	0.06	0.00	2.37	5.62	33	23	0	1	0	24	56
Chad	0.61	0.02	0.23	0.02	0.00	0.28	0.90	36	1	14	1	0	17	53
Congo, Rep.	1.02	0.53	0.04	0.51	0.00	1.08	2.10	62	32	2	31	0	65	128
Côte d'Ivoire	2.13	—	0.00	0.00	0.00	—	2.13	348	—	0	1	0	—	348
Congo, Dem. Rep.	0.00	0.00	0.06	0.00	0.00	0.06	0.06	0	0	4	0	0	4	4
Ethiopia	3.39	0.15	1.31	0.17	0.00	1.63	5.02	417	19	161	20	0	200	618
Ghana	1.20	0.56	0.21	0.55	0.05	1.36	2.56	129	60	22	59	6	146	275
Kenya	2.17	0.58	0.40	0.00	0.07	1.05	3.22	406	110	74	0	13	197	603
Lesotho	1.45	0.09	0.00	0.00	0.00	0.10	1.54	21	1	0	0	0	1	22
Madagascar	2.04	0.15	0.13	0.08	0.00	0.36	2.40	103	8	6	4	0	18	121
Malawi	1.65	0.51	0.04	0.00	0.00	0.56	2.21	47	15	1	0	0	16	63
Mali	1.77	0.36	0.19	0.23	0.38	1.16	2.93	94	19	10	12	20	61	155

Mozambique	0.96	—	0.89	0.07	0.01	—	0.96	63	—	58	5	1	63
Namibia	0.96	0.95	0.01	0.00	0.00	0.96	1.92	60	59	1	0	0	120
Niger	0.90	0.34	0.01	0.41	0.00	0.77	1.67	30	11	0	14	0	56
Nigeria	0.61	0.64	0.03	0.28	0.19	1.13	1.74	685	716	35	309	209	1,954
Rwanda	0.89	0.00	0.64	0.05	0.01	0.70	1.59	21	0	15	1	0	38
Senegal	1.61	0.07	0.21	0.14	0.18	0.60	2.21	140	6	18	12	16	192
South Africa	0.97	0.26	0.00	0.00	0.00	0.27	1.23	2,345	631	8	0	5	2,989
Tanzania	1.84	0.11	0.28	0.00	0.31	0.70	2.53	260	16	40	0	44	358
Uganda	1.19	1.38	1.01	0.00	0.75	3.14	4.33	104	121	88	0	65	378
Zambia	1.76	0.93	0.04	0.12	0.00	1.10	2.86	129	69	3	9	0	210
Middle-income	0.98	0.29	0.01	0.00	0.00	0.30	1.28	2,654	777	33	1	5	3,470
Resource-rich	0.73	0.56	0.03	0.33	0.13	1.05	1.78	1,629	1,240	75	736	278	3,959
Low-income, nonfragile	1.78	0.39	0.50	0.12	0.15	1.15	2.94	1,969	430	549	129	165	3,241
Low-income, fragile	1.49	0.00	0.10	0.55	0.03	0.68	2.16	571	0	37	210	12	830
Sub-Saharan Africa	1.10	0.37	0.11	0.17	0.07	0.72	1.81	7,033	2,370	694	1,076	460	11,633

Source: Authors.

Note: a. Average for 2001–05, except for Botswana, the Republic of Congo, and Mali, which are average for 2002–07.

ODA = official development assistance; OECD = Organisation for Economic Co-operation and Development; PPI = private participation in infrastructure; CAPEX = capital expenditure.

— = Not available.

Table A7.2   Size and Composition of the Power Sector Funding Gap[a]
current US$, million p.a.

	Total needs	Total spending	Total potential gain from achieved efficiency	Funding gap
Benin	(178)	114	17	**(47)**
Botswana	(116)	137	55	
Cameroon	(745)	258	343	**(145)**
Cape Verde	(24)	56	19	
Chad	(39)	53	66	
Congo, Rep.	(482)	128	45	**(310)**
Côte d'Ivoire	(825)	348	668	
Congo, Dem. Rep.	(1,473)	4	335	**(1,134)**
Ethiopia	(3,380)	618	82	**(2,681)**
Ghana	(728)	275	317	**(137)**
Kenya	(1,019)	603	109	**(306)**
Lesotho	(26)	22	20	
Madagascar	(478)	121	19	**(338)**
Malawi	(57)	63	91	
Mali	(178)	155	9	**(13)**
Mozambique	(771)	63	74	**(634)**
Namibia	(285)	120	0	**(166)**
Niger	(76)	56	126	
Nigeria	(7,736)	1,954	1,526	**(4,256)**
Rwanda	(118)	38	48	**(32)**
Senegal	(993)	192	231	**(570)**
South Africa	(13,511)	2,989	5	**(10,516)**
Tanzania	(910)	358	348	**(204)**
Uganda	(601)	378	158	**(64)**
Zambia	(472)	210	160	**(102)**
Middle-income	(14,191)	3,470	906	**(9,814)**
Resource-rich	(11,770)	3,959	3,541	**(4,269)**
Low-income, nonfragie	(9,704)	3,241	1,818	**(4,645)**
Low-income, fragile	(5,201)	830	2,021	**(2,350)**

Source: Authors.
Note: a. Average for 2001–05, except for Botswana, the Republic of Congo, and Mali, which are average for 2002–07.

Table A7.3 Sources of Potential Efficiency Gains, by Component[a]

	Current US$, million p.a.						GDP share (%)					
	Increase cost recovery	Improve system losses	Address under-collection	Reduce over-manning	Raise budget execution	Total	Increase cost recovery	Improve system losses	Address under-collection	Reduce over-manning	Raise budget execution	Total
Benin	1	6	0	10	0	17	0.01	0.15	0.00	0.23	0.01	0.40
Botswana	20	2	3	30	0	55	0.19	0.02	0.03	0.28	0.00	0.52
Cameroon	205	105	0	30	3	343	1.23	0.64	0.00	0.18	0.02	2.07
Cape Verde	1	3	8	6	1	19	0.09	0.31	0.78	0.60	0.08	1.85
Chad	43	17	0	6	0	66	0.73	0.29	0.00	0.10	0.01	1.13
Congo, Rep.	0	20	7	17	0	45	0.00	0.33	0.12	0.28	0.00	0.73
Côte d'Ivoire	—	—	626	42	0	668	—	—	3.83	0.26	0.00	4.09
Congo, Dem. Rep.	0	92	243	—	0	335	0.00	1.30	3.42	—	0.00	4.72
Ethiopia	42	24	15	—	0	82	0.34	0.20	0.13	—	0.00	0.66
Ghana	70	85	153	7	0	317	0.65	0.80	1.43	0.07	0.00	2.95
Kenya	8	54	0	31	15	109	0.05	0.29	0.00	0.17	0.08	0.58
Lesotho	15	5	0	—	0	20	1.05	0.32	0.03	—	0.01	1.41
Madagascar	0	19	0	—	0	19	0.00	0.37	0.00	—	0.00	0.37
Malawi	72	16	3	—	0	91	2.51	0.55	0.10	—	0.00	3.17
Mali	—	—	—	9	0	9	—	—	—	0.17	0.00	0.17

(continued next page)

Table A7.3 (continued)

	Current US$, million p.a.						GDP share (%)					
	Increase cost recovery	Improve system losses	Address under-collection	Reduce over-manning	Raise budget execution	Total	Increase cost recovery	Improve system losses	Address under-collection	Reduce over-manning	Raise budget execution	Total
Mozambique	36	21	0	18	0	74	0.54	0.31	0.00	0.27	0.00	1.12
Namibia	0	0	—	—	0	0	0.00	0.00	—	—	0.00	0.00
Niger	64	27	28	6	0	126	1.93	0.82	0.84	0.19	0.00	3.78
Nigeria	672	359	344	—	151	1,526	0.60	0.32	0.31	—	0.13	1.36
Rwanda	37	9	0	2	0	48	1.55	0.36	0.00	0.10	0.00	2.01
Senegal	165	51	0	14	2	231	1.90	0.58	0.00	0.16	0.02	2.66
South Africa	0	5	0	—	1	5	0.00	0.00	0.00	—	0.00	0.00
Tanzania	260	75	0	12	1	348	1.84	0.53	0.00	0.08	0.01	2.46
Uganda	0	64	86	8	0	158	0.00	0.73	0.98	0.09	0.00	1.81
Zambia	152	6	0	—	2	160	2.07	0.08	0.00	—	0.02	2.17
MIC	38	14	12	840	2	906	0.01	0.01	0.00	0.31	0.00	0.33
Resource-Rich	1,609	761	528	410	234	3,541	0.72	0.34	0.24	0.18	0.11	1.59
LIC-NoFragile	826	504	308	160	20	1,818	0.75	0.46	0.28	0.15	0.02	1.65
LIC-Fragile	0	498	1,423	99	0	2,021	0.00	1.30	3.71	0.26	0.00	5.26
SSA	2,323	1,340	1,842	1,147	210	6,862	0.36	0.21	0.29	0.18	0.03	1.07

Source: Authors.

Note: a. Average for 2001–05, except for Botswana, the Republic of Congo, and Mali, which are average for 2002–07.

Index

Boxes, figures, notes, and tables are indicated by *b*, *f*, *n*, and *t* following page numbers.